Sustainable Agriculture
and the Environment

Sustainable Agriculture and the Environment

Perspectives on Growth and Constraints

EDITED BY

Vernon W. Ruttan

Westview Press

BOULDER • SAN FRANCISCO • OXFORD

I am indebted to Sylvia Rosen, DeBora Burton, and Suzanne Denevan
for helping me transform an often incoherent transcript
of the consultation dialogues into a document that
could be submitted to Westview Press

This Westview softcover edition is printed on acid-free paper and bound in library-quality, coated covers that carry the highest rating of the National Association of State Textbook Administrators, in consultation with the Association of American Publishers and the Book Manufacturers' Institute.

Published in 1992 in the United States of America by Westview Press, Inc., 5500 Central Avenue, Boulder, Colorado 80301-2847, and in the United Kingdom by Westview Press, 36 Lonsdale Road, Summertown, Oxford OX2 7EW

Library of Congress Cataloging-in-Publication Data
A CIP catalog record for this book is available from the Library of Congress.
ISBN 0-8133-8507-5

Printed and bound in the United States of America

The paper used in this publication meets the requirements
of the American National Standard for Permanence of Paper
for Printed Library Materials Z39.48-1984.

10   9   8   7   6   5   4   3   2   1

Contents

Preface

This book is the product of one of a series of consultations held at the University of Minnesota Hubert H. Humphrey Institute of Public Affairs on November 27 and 28, 1989. The participants in the series were a number of leading biological and social scientists who were asked to identify the implications of global change for agricultural research priorities into the twenty-first century.

The idea of the consultations emerged out of a series of conversations in the spring of 1988 between Robert W. Herdt, Director for Agricultural Sciences at the Rockefeller Foundation, and the editor. The topic of the discussions was how to focus attention on emerging agricultural research priorities in developed and developing economies in response to the demands they would be placing on their agricultural systems and on their agricultural scientists as we move into the second quarter of the twenty-first century.

The consultations were organized around three broad subject areas: (a) Scientific and Technical Constraints on Crop and Animal Productivity; (b) Resource and Environmental Constraints in Sustainable Growth in Agricultural Production; and (c) Health Constraints in Agricultural Development. They were structured as informal dialogues. No formal papers were requested or presented. The dialogues were taped, and the transcripts were reproduced initially as Staff Papers at the University of Minnesota Department of Agricultural and Applied Economics.

This publication is based on the second of the series. Copies of the first and third productions can be obtained from Waite Library, Department of Agricultural and Applied Economics, University of Minnesota, 1994 Buford Avenue, St. Paul, MN 55108.

Vernon W. Ruttan

About the Contributors

Dean E. Abrahamson is professor of public affairs at the Hubert H. Humphrey Institute of Public Affairs, co-chair of the All-University Council on Environmental Quality at the University of Minnesota, and visiting professor at the Department of Environmental and energy Systems, Lund University, Lund, Sweden. He is a trustee of the Natural Resources Defense Council and a fellow of the American Association for the Advancement of Science and of the Scientists Institute for Public Information. His primary research interest is in the intersection of energy and environmental policies, particularly the implications of a commitment to nuclear power.

C. Eugene Allen is vice president with responsibility for Forestry and Home Economics at the University of Minnesota. He has been dean of the College of Agriculture and associate director of the Experiment Station at the University of Minnesota since 1984. Previously, he was a faculty member in the departments of Animal Science and Food Science and Nutrition. His research on animal growth biology and the functional and nutritional characteristics of animal food products has earned him numerous awards from professional societies. He presently serves on the Board on Agriculture of the National Research Council of the National Academy of Sciences.

Zbigniew Bochniarz is visiting associate professor in the Hubert H. Humphrey Institute of Public Affairs, University of Minnesota. He is on leave from the Warsaw (Poland) School of Economics. His research interests are in the fields of environmental problems and sustainable development of economic reforms in Central and Eastern Europe. His published works include *Designing Institutions for Sustainable Development: A New Challenge for Poland* (1991) and *The Economic Problems of Environmental Protection* (1988).

Robert Chen, geographer and climatologist, is currently assistant professor (Research) with the Alan Shawn Feinstein World Hunger Program at Brown University. He has held research fellowships with the Environment Program of the International Institute for Applied Systems Analysis in Austria and with the Board on Atmospheric Sciences and Climate at the U.S. National Academy of Sciences in Washington, D.C. He is co-editor of *Social Science Research and Climate Change* (with S. Schneider and E. Boulding) and has worked actively in the field of climate impact assessment for more than 15 years. At the World Hunger Program, his research has focused on the ability to meet future food needs in the light of possible global environmental change, on measures of the prevalence of hunger, and on hunger among refugees. Dr. Chen also has contributed to research projects on the history of hunger, surprising African futures, assessment of vulnerability to famine, and new initiatives for overcoming hunger.

H. H. Cheng, professor and head of the Department of Soil Science at the University of Minnesota, formerly was a faculty member at Washington State University for 24 years. His research interest is in soil biochemistry and analytical chemistry. He has conducted research on the fate and transformations of natural and synthetic organic materials and chemicals in the environment. He is editor of the Soil Science Society of America monograph, *Pesticides in the Soil Environment: Processes, Impacts, and Modeling* (1990). Dr. Cheng is a fellow of the Soil Science of America (1983), the American Society of Agronomy (1983), and the American Association for the Advancement of Science (1990).

William C. Clark is assistant director of the Center for Science and International Affairs in the John F. Kennedy School of Government at Harvard University. Before joining the Kennedy School in 1987, he led the team of scholars from eastern and western countries in the Program on Sustainable Development of the Biosphere at the International Institute for Applied Systems Analysis in Austria. He has also worked with Alvin Weinberg at the Institute for Energy Analysis in Oak Ridge and with C. S. Holling at the Institute of Resource Ecology in Vancouver. His current research focuses on policy issues arising through the interactions of environment, development, and security concerns in international affairs. He is a member of the U.S. National Academy of Sciences' Committee for Global Change, and he serves on the steering committees for the U.S. Office of Interdisciplinary Earth Studies, Resources for the Future's Climate Program, and the International Institute for Applied Systems Analysis's work on environmentally sound restructuring of national economies. He is co-editor of *Environment*, a monthly magazine of international environmental affairs. In 1983, Clark was awarded a MacArthur Prize Fellowship for his achievements in environmental policy.

Pierre Crosson is an economist and senior fellow in the Climate Resources Program, Resources for the Future, where he has been employed for over 25 years. His research has dealt with a range of issues concerning the relations between agricultural production, natural resources, and the environment. He has given special attention to the costs of erosion-induced losses of soil productivity and of downstream damages of sediment and related chemical pollutants.

Margaret Bryan Davis is Regents' Professor of Ecology in the Department of Ecology, Evolution and Behavior at the University of Minnesota. She served as department head, 1976-1981. Her research has been in the field of forest ecology, vegetation response to Quaternary climatic change, long-term ecosystem processes, watershed-lake interactions, sedimentation in lakes, and expected ecosystem responses to future Greenhouse Warming. She served as president of the American Quaternary Association and the Ecological Society of America and is a fellow of the Geological Society of America (1966), the American Association for the Advancement of Science (1968), the American Academy of Arts and Sciences (1991), and a member of the National Academy of Sciences (1982). She received the Award for Distinguished Contributions to the Advancement of Science and Technology from the Science Museum of Minnesota, the Alumnae Recognition Award from Radcliffe College in 1988, and the Award for Women in Academic Science by the Minnesota Women's Consortium in 1991.

Robert W. Herdt, director for Agricultural Sciences at the Rockefeller Foundation since 1987, has actively worked with biological and social scientists over the past two decades to improve production and income of small-scale farmers in developing countries. He has headed the Economics Department of the International Rice Research Institute, acted as Scientific Advisor to the CGIAR Secretariat at the World Bank, and wrote *The Rice Economy* (with Randolph Barker) and *Science* *and Food*: *The CGIAR and Its Partners* (with Jock Anderson and Grant Scobie).

Richard Jones is dean of the College of Agriculture at the University of Minnesota. He has been a faculty member in the Department of Entomology since 1977 and head of the Department of Entomology (1984). For six years he directed the teaching and research mission of that major college department prior to serving as interim associate director for the Agricultural Experiment Station and to his selection as dean. His expertise in insect physiology and behavioral chemicals is nationally and internationally recognized. He was selected as a member of a three-person Office of International Cooperation and Development team to survey and evaluate the use of biological control for stem-boring insects in the People's Republic of China. Since 1985, he has served as adviser to a PL 480 project in Yugoslavia on the international management of corn insects. He has received many state and federal grants to investigate the biological control of insects and has authored many articles and book chapters on chemical pest management. Dr. Jones is co-author of *Semiochemicals*: *Their Role in Pest Control*.

William E. Larson is Emeritus Professor and former head of the Soil Science Department at the University of Minnesota. His field of interest is soil management, tillage, and erosion. He is a fellow of the Soil Science Society of America, the American Society of Agronomy, the Soil and Water Conservation Society, and the American Association for the Advancement of Science. He received the Soil Science Research Award in 1973, and the Merit Award and an honorary membership from the World Association of Soil and Water Conservation in 1989. He is a past president of the Soil Science Society of America and the American Society of Agronomy.

Robert D. Munson, vice president for Research, Education and Market Development for the National Fertilizer Solutions Association and Fluid Fertilizer Foundation, was formerly project associate at the Center for International Food and Agricultural Policy and adjunct professor in the Soil Science Department of the University of Minnesota and a private consultant. His research interest has been in soil fertility and plant nutrition, particularly potassium, soil testing and plant analysis, the interrelation of essential elements, and the interactions of cultural and management practices that increase crop yields, as well as the economics of fertilizer use. More recently his focus has been on improving nutrient use efficiency to improve agricultural competitiveness and environmental quality. He is editor of *Potassium in Agriculture* published by the American Society of Agronomy, Crop Science Society of America and the Soil Science Society of America. He is a fellow in the American Association for the Advancement of Science (1963), American Society for Agronomy (1974), Soil Science Society of America (1976), and Crop Science Society of America (1985).

Stephen L. Rawlins has been national program leader for Soil Erosion/Global Change since 1988 at the U. S. Department of Agriculture, Agricultural Research Service. He has been associated with the Agricultural Research Service since 1980: first as assistant to the deputy administrator in the alternative energy strategies program for agriculture (1980-81); as national program leader, Tillage, in the program to develop strategies to conserve water and soil (1981-83); and as national program director of ARS, directing overall research strategies for natural resource management. From 1985-88, he was at the Systems Research Laboratory, Beltsville, where he used systems engineering to package scientific knowledge into computer-aided decision support systems to solve important agricultural problems, developing soil environment modules for modularized crop simulator.

Steve Rayner, deputy director of the Global Environmental Studies Center at Oak Ridge National Laboratory, Tennessee, leads interdisciplinary research programs that combine social and natural sciences. Since 1980 he has worked in the United States in the field of risk perception, communications, and management. He has published extensively on the different concepts of nature and of fairness that institutions invoke in decision making about technology and the environment. He was principal author of a recent report to the U.S. Congress on policies to encourage private sector response to climate change. Also, he has advised governmental and scholarly institutions on the human dimensions of global change, including the U.S. Department of Energy, the Agency for International Development, the Office of Technology Assessment, the Consortium for International Earth Science Information Network, the International Social Science Council, and the U.S. Social Science Research Council.

Norman J. Rosenberg is a senior fellow and the director of the Climate Resources Program, Resources for the Future. Previously, he was George Holmes Professor of Agricultural Meteorology and director for Agricultural Meteorology and Climatology at the University of Nebraska-Lincoln. He is editor of *Drought in the Great Plains: Research on Impacts and Strategies* and author of *Microclimate: The Biological Environment*, and a number of papers on the issue of climatic change and its possible effects on agriculture and natural ecosystems. Recently he was on the National Academy of Sciences/National Research Council's panel on Policy Implications of Climate Change. He is also co-chair of a U.S. Department of Agriculture-sponsored study on Implications of Climate Change for U.S. Agriculture and Forestry, Council for Agricultural Science and Technology (CAST).

Vernon W. Ruttan is Regents Professor in the Department of Agricultural and Applied Economics and Department of Economics, and adjunct professor, Hubert H. Humphrey Institute of Public Affairs, University of Minnesota. From 1965-1970 he was professor and head of the Department of Agricultural and Applied Economics, and from 1970-1973, director of the Economic Development Center. He has been a visiting professor at the University of California, Berkeley (1958-59) and University of Philippines (1963-65). His non-academic experience includes the staff of the President's Council of Economic Advisors (1961-62), agricultural economist with the Rockefeller Foundation at the International Rice Research Institute in the Philippines (1963-65), president of the Agricultural Development Council (1973-78), and a number of Advisory Committees and Boards. He was president of the American Agricultural Economics Association in 1971-72. His research focus has been the economies of technical change and agricultural development. His publications include *Agricultural Development: An International Perspective* (with Yujiro Hayami), *Agricultural Research Policy, Aid and Development* (with A. O. Krueger and C. Michalopoulos), and many journal articles and book chapters. He received six awards for published research from the American Agricultural Economics Association. He also is a fellow of the American Agricultural Economics Association, the American Academy of Arts and Sciences, American Association for the Advancement of Science, and a member of the National Academy of Sciences. He holds honorary degrees from Rutgers University, Purdue University, and the Christian Albrechts University of Kiel. He has received the Alexander von Humboldt Award for outstanding contribution to agriculture and a Distinguished Service Award from the U.S. Department of Agriculture.

Pedro A. Sanchez is professor of Soil Science and coordinator of

North Carolina State University's Tropical Soils Research Program and adjunct professor of tropical conservation at Duke University. A native of Cuba, he joined the N. C. State faculty in 1968. His professional career has been dedicated to improving the management of tropical soils for sustained food production and protection of the natural resource base. Sanchez has been stationed in the Philippines, Peru (twice), and Colombia, where he worked on rice research, pasture research, research administration, and institutional development at national institutions and international agricultural research centers. Since 1972 he has been coordinator of N. C. State's Tropical Soils Research Program with principal research locations in Peru, Brazil, and Madagascar and network operations throughout tropical America and Africa. He currently serves as board chairman of the world-wide Tropical soil Biology and Fertility Program, as Chairman of a National Academy of Sciences panel on Sustainable Agriculture and the Environment in the Humid Tropics, and as director of the Center for World Environment and Sustainable Development of Duke University, North Carolina State University, and the University of North Carolina at Chapel Hill. His publications include *Properties and Management of Soils of the Tropics*, seven edited books, and over 10 scientific articles written in English, Spanish, and Portuguese. He is a fellow of the American Society of Agronomy and the Soil Science Society of America. He has been decorated by the governments of Colombia and Peru. Effective October 1, 1991, he will be director general, International Council for Research in Agroforestry, Nairobi, Kenya.

Steven Sonka conducts teaching and research on strategic change and decision making in the food and agribusiness management sector. He has authored or co-authored more than 125 books, articles, and publications. He is best known for his unusual ability to relate his scholarship to practical issues of importance to decision makers. His major areas of research include use of information technology in business decision making, evaluation of strategic marketing alternatives within agriculture, and the role of climate in agribusiness decision making. Dr. Sonka teaches undergraduate and graduate courses in business management of farm and agribusiness firms and use of decision support systems in agriculture. He has served as consultant to special projects of the Organization for Economic Cooperation and Development in Paris, to special task forces in the U. S. Senate, and to the Office of Technology Assessment of the U.S. Congress. His international experiences include consulting and lecturing in Australia, Brazil,

England, France, Japan, and Spain.

Paul E. Waggoner is a meteorologist and plant pathologist. From 1951-1987 he worked at the Connecticut Agricultural Experiment Station, New Haven, as a scientist and director. He is retired but continues at the Station as distinguished scientist. His research spans physical phenomena in agriculture from plant diseases and microclimate to forestry and water resources and recently has concerned adaptation to climate change.

PART ONE

Introduction

1

Concerns About Resources
and the Environment

Vernon W. Ruttan

In this book a group of leading agricultural and social scientists explore the resource and environmental constraints on sustainable growth in agricultural production into the middle of the twenty-first century. Contemporary concerns with the implications of natural resource availability and environmental change are presented in this section to provide the broader historical context.

We are now in the midst of the third wave of social concern since World War II with the implications of natural resource availability and environmental change for the sustainability of improvements in human well-being.

The *first* wave, in the late 1940s and early 1950s, focused primarily on the quantitative relations between resource availability and economic growth--the adequacy of land, water, energy, and other natural resources to sustain growth. The reports of the U.S. President's Water Resources Policy Commission (1950) and the U.S. President's Materials Policy Commission (1952) were landmarks of the early postwar resource assessment studies generated by this first wave of concern. The primary response was technical change. In retrospect, it appears that a stretch of high prices has not yet failed to induce the new knowledge and new technologies needed to locate new deposits, promote substitution, and enhance productivity. If the Materials Policy Commission were writing today, it would undoubtedly have to conclude that there has been abundant evidence of the nonevident becoming the evident; the expensive, cheap; and the inaccessible, accessible (Ausubel and Sladovich, 1989; Barnett and Morse, 1963).

The *second* wave of concern occurred in the late 1960s and early 1970s. The earlier concern with potential "limits to growth" imposed by natural resource scarcity was supplemented by concern with the capacity of the environment to assimilate the multiple forms of pollution generated by growth. An intense conflict was emerging between two major sources of demand for environmental services. The first was the rising demand for environmental assimilations of residuals derived from growth in commodity production and consumption (e.g. asbestos in our insulation, pesticides in our food, smog in the air, and radioactive wastes in the biosphere). The second was the rapid growth in per capita income and high income elasticity of demand for such environmental services as access to natural environments and freedom from pollution and congestion (Ruttan, 1971). The response to these concerns--still incomplete--was the organization of local institutions designed to force individual firms and other organizations to bear the costs rising from the externalities generated by commodity production.

Since the mid-1980s, these two earlier concerns have been supplemented by a *third*. The newer concerns center on the implications for environmental quality, food production, and human health of a series of environmental changes that are occurring on a transnational scale: issues such as global warming, ozone depletion, acid rain, and others (Committee on Global Change, 1990; Committee on Science, Engineering, and Public Policy, 1991). The institutional innovations needed to respond to these concerns will be more difficult to design. Like the sources of change, they will need to be transnational or international. Experience with attempts to design incentive-compatible transnational regimes, such as the Law of the Sea Convention or even the somewhat more successful Montreal Protocol on reduction of CFC emissions, suggests that the difficulty of resolving free rider and distributional equity issues imposes a severe constraint on how rapidly the efforts of transnational regimes to resolve these new environmental concerns can be put in place (Dorfman, 1991).

It is of interest that, with each new wave of concern, the issues that dominated the earlier wave were recycled. The result is that while the intensity of earlier concerns has receded, in part due to the induced technical and institutional changes, the concerns with the relations between resource and environmental changes and sustainable growth in agricultural production have broadened. During the 1980s, for example, concerns with the effects of more intensive agricultural production (a) on resource degradation through erosion, salinization and depletion of groundwater; and (b) on the quality of surface and groundwater through runoff and leaching of plant nutrients and pesticides, intensified. Terms that initially had been introduced by the populist critics of agricultural research, such as alternative, low input, regenerative, and sustainable agriculture, began to enter the vocabulary of the people responsible for agricultural research resource allocation.

The Agricultural Transformation

Contemporary concerns about the capacity to sustain the growth in the demands that societies are placing on their agricultural systems have emerged during the completion of one of the most remarkable transitions in the history of agriculture. Prior to this century almost all increases in food production were obtained by bringing new land into production. The few exceptions to this generalization were in limited areas of East Asia, the Middle East, and Western Europe (Hayami and Ruttan, 1985).

By the first decade of the twenty-first century, almost all increases in world food production will have to come from higher yields: increased output per hectare. In most of the world the transition from a resource-based to a science-based system of agriculture is occurring within a single century. In a few countries this transition began in the nineteenth century. Wheat prices, corrected for inflation, have declined since the middle of the nineteenth century. Rice prices have declined sine the middle of the twentieth century. These trends suggest that productivity growth has been able to more than compensate for the rapid growth in demand, particularly during the decades since World War II.

As we look toward the future, however, the sources of productivity growth are not as apparent as they were a quarter century ago. The demands that the developing economies will place on their agricultural producers owing to population growth and growth in per capita consumption arising out of higher income will be exceedingly high. Population growth rates are expected to decline substantially in most countries during the first quarter of the twenty-first century. But the absolute increases in population size will be large and increases in per capita income will add substantially to food demand. The effect of growth in per capita income will be much more rapid growth in demand for animal proteins as well as for maize and other feed crops. During the next several decades growth in food and feed demand arising from growth in population and income will run upwards of 4.0 percent per year in many countries. Many countries will experience more than a doubling of food demand before the end of the second decade of the twenty-first century.

It is apparent that the gains in agricultural production required over the next quarter century will be achieved with much greater difficulty than in the immediate past. Currently, difficulty is being experienced in raising yield ceilings for the cereal crops that had rapid yield gains in the recent past (Ruttan, 1989). The incremental response to increases in fertilizer has declined. Expansion of irrigated area has become more costly. Maintenance research, the research required to prevent yields from declining, is rising as a share of research efforts. The institutional capacity to respond to these concerns is limited, even in countries with the most effective national research and extension systems. Indeed, there has been considerable difficulty in many countries during the 1980s in maintaining the agricultural research capacity that was established during the 1960s and 1970s.

It is possible that within another decade advances in basic knowledge will create new opportunities for advancing agricultural technology to reverse the urgency of some of the preceding concerns. Institutionalization of private sector agricultural research capacity in some developing countries is beginning to complement public sector capacity. Advances in molecular biology and genetic engineering are occurring rapidly. But the date when these promising advances will be translated into productive technology seems to be receding. Thus it is particularly appropriate at this time to examine the implications for the international agricultural research agenda of the constraints which resources and the environment may impose on the sustainable growth of agricultural production into the twenty-first century.

The conversations reproduced in this book are organized around four general topics. The first focuses on what we really know about global climate change. What can we actually say we know compared to what we think or hope? In this group we have Dean Abrahamson, Norman Rosenberg, William Clark, and Steve Sonka. Abrahamson starts the discussion.

The second topic centers on the impact of global climate changes on agriculture and natural resources. Steve Rawlins provides the transition between the first and second groups, and then we have Paul Waggoner, Margaret Davis, Robert Chen, Zbigniew Bochniarz, and Pierre Crosson participating in the conversation. I visualize Crosson's presentation as the transition to the issues of micro-environmental change. Pedro Sanchez discusses some of the problems associated with tropical soils and tropical ecologies, and William Larson and H. H. Cheng discuss temperate region soil problems. Richard Jones discusses pests and pathogens that arise from both global climate change and some micro-environmental change, and Robert Munson reflects on some of his concerns with soil fertility.

Finally, we move to a discussion of the implications for agricultural research. Steve Rayner is concerned with decision making under the kinds of uncertainty that arise from the changes under discussion, and Steve Rawlins reflects on the research implications from a federal perspective. Eugene Allen discusses research implications from a state perspective.

It is from the open exchange of ideas from a variety of perspectives that important ideas arise and are defined. It is my hope that the readers of this book will continue the conversations and advance the programs needed to achieve sustained increases in agricultural production. They will be needed as we move into the first decades of the twenty-first century.

PART TWO

Global Climate Change

2

Greenhouse Gases and Climate Change

Dean E. Abrahamson

Abrahamson: At current growth rates in the emissions of greenhouse gases, we can expect increases in concentrations equivalent to a doubling of the pre-industrial concentration of CO_2 (carbon dioxide) by about 2030. Most of the general circulation models show an equilibrium heating -- an increase of annual average global temperature -- in the neighborhood of $3°$-$5°$ Celsius for a doubling of CO_2 concentration. Because of the large heat capacity of the oceans, and the time to transfer heat into the oceans, there is an ocean thermal delay generally represented as 20-40 years. That is the reason for the lag between the atmospheric concentration of greenhouse gases and the observed increase in atmospheric temperature.

A number of biogeochemical feedbacks are not incorporated into the general circulation models. They are not well understood. One estimate by Lashoff (1989) suggests that taking these feedbacks into account could lead to a global heating equivalent to that shown by the general circulation models. If that should be the case, and if we permit emission rates to continue at their present levels, we could be committed to a warming of 6-$8°$ within the next three or four decades. Some of these feedbacks have a long-time constant. Some don't.

The global heating -- the concentration of greenhouse gases and the resultant climate change -- is irreversible: not, of course, in geologic time frames but in social or economic or political relevance time. The issue differs from most conventional environmental issues in that there are, for practical purposes, no scrubbers -- there is no simple technical fix. There are some exceptions: chlorofluorocarbons and some of the other industrial chemicals. This means that in order to reduce emission rates the level of activities that produce these gases must be reduced.

Finally, I find it hard to convince myself that we aren't already committed to a warming of 3-5° Celsius as an equilibrium warming because of gases already in the atmosphere, plus those certain to be released while we debate policy response and implement it.

What about uncertainty? The basic physics of the greenhouse effect is very well known. That there will be climatic change resulting from greenhouse gas emissions is a certainty. How much and when is not at all certain.

We started last spring to put together a faculty seminar at the University that will consider implications of climatic change for Minnesota. It is the first draft of a scenario for climatic change for Minnesota but has not yet been reviewed. It was done by my colleague, Peter Ciborowski.

We were deliberately conservative. We didn't take the high end of the scale for temperature or precipitation. The base case suggests an average temperature increase of one degree by the turn of the century and three degrees by 2030; additional frost-free days by 2030; a decline in degree heating days; an increase in degree cooling days; a 2 or 3 week earlier snow melt; 10 or 20 additional July days over 90°F; a reduction in summer soil moisture of 25 percent; and increased drought frequency and decreased runoff. In my view the changes are apt to be much more extreme, but I tend to be a little more pessimistic than some people.

In 1987 a conference sponsored by the World Meteorological Organization and UNEP and a couple of other organizations was held in Villach, Austria. Several of you were there. The group put together a scenario that projected larger temperature increases at high latitudes than at low, and more in winter than in summer.

The National Academy of Sciences, in a review published in 1987, addressed the question of likelihood. Large stratospheric cooling was regarded as virtually certain and global mean surface warming very probable. The suggested range for an equivalent doubling of CO_2 is 1-1/2° to 4-1/2° Celsius. Global mean precipitation increase is very probable as are the reduction of sea ice, polar winter surface warming, and rise in sea level. Summer continental dryness and warming are likely in the long term. The effects of climatic change are probably nonlinear. Some of the physical impact, sea level rise, for example, may be a linear function of average warming over some range, but the impact of the sea level rise will be highly nonlinear. I can't think of any impact that scales linearly with temperature.

The policy situation is evolving very rapidly. You don't hear talk anymore about winners and losers or that society's only response can be adapting to climatic change. You did a few years ago, but the talk pretty much disappeared as soon as the effects associated with rates of climatic change were understood.

Clearly, the emissions of greenhouse gases have to be reduced, and reduced substantially. Various estimates have been made of the necessary reductions in emissions of greenhouse gases to stabilize their atmospheric concentrations. For example, the EPA (Lashof & Tirpak, 1989) suggested a reduction in CO_2 emissions of between 50-80 percent; a 10-20 percent reduction in methane; elimination of the long life CFCs, and reduction of nitrous oxide of, as I recall, 85 percent. Thus, very, very substantial reductions in these gases are required to stabilize atmospheric greenhouse gas concentrations.

Bills pending before the Congress and other legislative bodies range all the way from stabilizing emissions to reductions of 20 percent or so by the end of the century. About 60 percent of the greenhouse gases appear to come from fossil fuels, the rest from a number of other sources. At the same time that we will be trying to restructure energy production and use to reduce emissions it will be necessary to try to cope with substantial unavoidable climatic change. Both activities are potentially disruptive: socially, economically, politically, and ecologically. And both are very expensive.

Crosson: The report Martin Parry put together for the International Institute for Applied Systems Analysis (Parry, 1990) talks about winners and losers as far as agriculture is concerned. The argument is that after reviewing all the material there would be no reason to assume much effect on global agricultural capacity with an equivalent doubling of CO_2. Clearly some regions would lose but others would benefit. It is a kind of standoff as far as global agricultural capacity is concerned.

Abrahamson: If you consider local effects only, everybody won't come out equally. Minnesota won't fare the same as Kansas, for example. If you postulate some new equilibrium climate and we move into it slowly, then you can conjure up credible winner scenarios. If the rates of climatic change are as high as they now appear, then it is difficult to imagine that the disruption associated with these high rates of change would be anything but serious.

Rayner: I clearly move in different circles from Abrahamson because I hear the issue of winners and losers discussed quite extensively. The issue is seldom thought through adequately. Certainly, in the short term winners and losers are likely. There is considerable uncertainty about when the doubling equivalent would be reached. Abrahamson mentioned 2030. I've heard dates that put it much further back in the next century. And, of course, there's also uncertainty attached to the ocean water buffering effect. What is meant by short term and long term varies. But more to the point, we should stop thinking about winners and losers in absolute terms. There may be no absolute winners over the present situation but some people are going to lose more than others.

Furthermore, we have to bear in mind that there are going to be winners and losers from preventive policies as well as from adaptive strategies. The cost of prevention may prove to be extremely high. EPA's estimates for protection of the coast line from sea level rise, for example, are impressive, but it is quite moderate by comparison with the costs of completely reconfiguring the U.S. energy system. So there will be winners and losers in the sense that large transfers of wealth will be involved in whatever response we make to climate change -- whether it's essentially preventive or adaptive. It's worth bearing in mind that in net terms, some are going to be better off and some worse off whatever we do.

Rawlins: In both the Global Change Working Group of the Committee on Earth Sciences and in the Impacts Assessment Panel of the IPCC, where I've been serving as a USDA representative, there is debate about winners and losers. But there is considerably more debate about the credibility of the projections for global warming. The report, *Scientific Perspectives on the Greenhouse Problem* (George C. Marshall Institute, 1989), is one example of the critical look now being taken at the scientific underpinning of these projections. The earth is a very complex system. The simplistic models on which global warming projections are being made are not capable of taking into account all the complex interactions that could influence the results.

We know with certainty that the atmospheric concentrations of carbon dioxide, methane, and nitrous oxides have increased, and that they will, most likely, continue to increase for some time to come. The consequences of these increases will be less certain. Our simplistic models predict that the physical consequence of the increased concentrations acting alone should be global warming. The question is, are they acting alone?

To illustrate how little we know about the complex earth system, consider the carbon cycle. We've lost about 130 teragrams of carbon from fossil reserves and from cement manufacture and another 150 to 160 teragrams from land use changes, including deforestation. The atmosphere has gained only 60 teragrams. The remainder went somewhere. Obviously the ocean is a big sink. The only other sink is the biosphere. Recently, oceanographers have decreased their estimates of the sink capacity of the ocean, leaving a large quantity of carbon unaccounted for. Could it be that the 25% increase in atmospheric carbon dioxide concentration we have already experienced has increased the biosphere capacity to fix carbon? We simply do not know. Both the ocean's and the biosphere's capacity to fix carbon are dependent upon complex processes that are affected by temperature and carbon dioxide concentration. The amount of carbon that ends up in the atmosphere is a small residual left over from some very large and dynamic processes that we're only beginning to understand. The error in estimating this residual could be huge.

I hear a lot more debate about winners and losers and about uncertainties of our projections now than I did six months or a year ago.

Ruttan: How strongly does the rate of change that Abrahamson emphasized affect the conclusions that we draw?

Rawlins: The less the rate of change the easier it will be to adapt to. We should not, however, assume that all changes will have negative impacts. The direct effect of increased carbon dioxide concentration on crop production, taken by itself, could be positive. Martin Parry's draft report referred to by Crosson takes this into account.

Rayner: For both forestry and agriculture?

Rawlins: Only agriculture was considered.

Rayner: But that's highly uncertain because all the experiments in CO_2 fertilization -- and we've done a lot at Oak Ridge -- are being done under very artificial circumstances.

Rawlins: Some experiments being conducted by ARS in cooperation with DOE now use a free air release of carbon dioxide in an open field.

Rayner: Well, they haven't all been completely enclosed, but they've been small plots about the size of this room. They're quite artificial, particularly from the point of view of the effects of pests.

Rawlins: I certainly agree with that.

Rayner: The importance of the impact of pests on CO_2-fertilized plants is that the pests must eat a lot more plants to get the same amount of nutrition. The whole thing is very up in the air as to whether there would be any benefit. We just don't know.

Rawlins: I agree. We don't know. Pest interactions must be taken into account. We're studying pest effects whether we want to or not in the FACE (Free Air Carbon Exchange) experiment being conducted by DOE, Brookhaven National Laboratory, ARS, and the University of Arizona at Maricopa. To overcome the problems of chamber walls carbon dioxide is released from standpipes encircling 20-meter-diameter plots. But leaf temperatures of the high carbon dioxide plots run higher than the surrounding plants, which selectively attracts insects.

Ruttan: We're going to come back to this issue again. Abrahamson, do you have any comment before we move to Rosenberg?

Abrahamson: Just a couple of things: One is that future emissions are under our control. We can't continue present trends. The other is the time perspective. I consider short-term to be a few hundred years. And I am not impressed with arguments about time periods shorter than a couple decades.

Rayner: The problem, though, is that the policy decisions are made by institutions today that don't have time frames of several hundred years. We are stuck with their time frames insofar as decision making on those issues is concerned.

Abrahamson: I appreciate that fully. Although I'm in the School of Public Affairs and supposed to understand how these institutions work, I try to forget from time to time.

Chen: Your base case is for a changed climate: Is there a base case for an unchanged climate?

Abrahamson: I don't think it's likely. We must do everything we can to reduce the rate of emissions -- to move away from the fossil fuels as rapidly as possible. I don't care what the cost is: the cost of not doing anything is, in my view, so large that we simply don't have that option. We're going to have to cope with whatever climatic changes are already in the mill. If we're lucky, and if all uncertainties come out at the low end, and we're vigorous in terms of reducing emissions, we might limit the change to a couple of degrees.

Davis: I want to respond to Dr. Chen's questions about a continuation of the present climate. The present climate is variable. We don't know whether the changes we've seen since the beginning of the century are actually an incipient greenhouse effect or a natural trend. But there is something to be learned from the past record. The rate of warming from 1900 to 1950 was rather similar to the lowest rate of possible response to increased CO_2. We only need to look at the response of agriculture to the series of droughts during the 1930s, 1940s, and 1950s to give us some idea of the best possible case scenario.

3

Climate Change Models

Norman J. Rosenberg

Rosenberg: That leads in very nicely to what I thought I would salvage from my little presentation because Rawlins covered much of what I wanted to say. Let me respond to some of Abrahamson's comments. I agree that if the current rate of emissions of greenhouse gases continues, there will be a significant impact on the global climate system. I'm that convinced of the physics of the process. But I don't have much faith in much of the detail that's available to us right now. I'm very skeptical of the kinds of numbers Abrahamson used in his scenario.

They say that there are two things that one shouldn't watch being made: laws and sausages. Global climate models are a third. The modeling effort is intellectually stimulating but very, very flawed. The kinds of numbers that are being quoted are very shaky. For example, the GFDL model that predicts the extreme heat and dryness in the mid-continent of North America produces radically different results from the NCAR climate model, largely because of the way the two parameterize soil moisture in the spring. Just a simple thing like whether the soil is full of water creates a tremendous difference in the outcome of the models.

Lashoff's analysis indicates that if all the feedbacks are positive, then the greenhouse effect will be much more severe than even the GCM modelers predict. But, on the other hand, some negative feedbacks are possible, too, and we just don't know which ones are going to play out.

The models need to be improved and the people working on the models understand this full well. It's the people -- often on the policy side, on the advocacy side, and in the media -- who grab the worst case and run with it. After all, it's much more interesting to know that the Midwest is going to turn into a permanent desert than it is to speculate about half a degree increase or decrease in temperature. The scientific basis on which so many of these projections are being made is still very weak. We have to be very skeptical and to remain skeptical.

Clearly, one of the few things we know for sure is that plants respond to elevated levels of carbon dioxide. There is a greater rate of photosynthesis and an effect on the stomatal functioning that causes plants to use less water. Rayner is right, of course, that most studies have been done in laboratories and growth chambers. What will happen at a field level outdoors is uncertain, but something will happen. The point that Rawlins made -- that only 60 gigatons have accumulated in the atmosphere while 300 have been emitted -- is relevant. The oceanographers are now becoming convinced that the capacity of the oceans to absorb carbon dioxide is much more limited than they had originally thought. That tells me that, since the CO_2 is not going to outer space, it's going into the terrestrial biosphere. But we don't know whether it's into forests or tundra or peat bogs or into that one percent per annum yield increase that the plant breeders take credit for. I suspect that the CO_2 "fertilization effect" is built into the yield increase which we've seen over the last 50 or 60 years.

Another important point is the rate of change. Surely, if the rate of change is very rapid, we're going to have more problems than if it is slow. That would be true for agriculture, for forests, water resources, and everything else that worries us. The temperature records produced by the group at East Anglia (Jones, Raper, et al. 1986) and at the Goddard Institute for Space Studies (Hansen & Lebedeff 1987) show that a net change in mean global temperature of slightly more than half a degree Celsius has occurred over the last 100 years (Figure 1). A portion of that is probably due to the urban "heat island effect," so the real number is probably smaller.

This temperature rise could be the result of normal climate variability. It could be an indication of a greenhouse effect. I find it rather upsetting when scientists grab hold of one side of an issue and quote only that which seems to support their argument. Lawyers are paid to be advocates and they try to make sure that only the information supporting their case reaches the judge and jury. However, what is laudable behavior in an advocate is not necessarily laudable in a scientist. Now I believe that the temperature record is essentially correct: that there has been a global warming trend over the last 100 years. But when we see that data for the continental United States show no change in temperature over exactly the same period of time, it should cause some doubt. The continental United States has the best

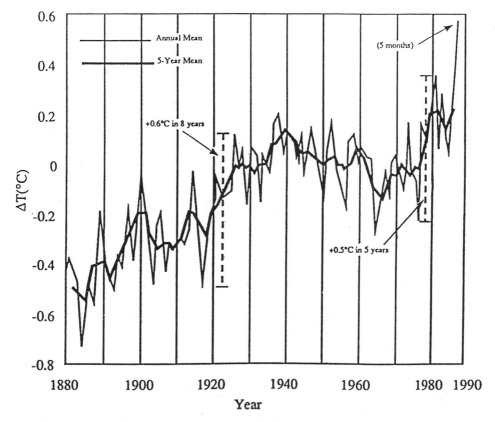

Figure 1. Global temperature trend 1880-1990. Adapted from J.I. Hanson, A. Furg, S. Lebedeff, D. Rind, R. Ruedz, G. Russell. Prediction of near term climates evolution: What can we tell decision-makers now? In J. Topping (Ed.) *Proceedings of the first North American conference on preparing for climate change* (Washington, D.C.: Climate Institute, 1987).

geographical distribution of temperature and precipitation measurement stations anywhere in the world. No change in temperature and no change in precipitation are shown in the record.

The counterargument is that the United States is only one small portion of the world. But it is the portion of the earth's surface where "greenhouse warming" is predicted, by some GCMs, to be most apparent. Yet we don't find it. We find cooling in many places in the world where we expected warming, and we find the greatest warming in the tropics where no warming was expected. My point is that the state of the science is really far too weak to support the kinds of major policy decisions that are being proposed right now by many advocates.

A number of analysts at RFF are engaged in a study, with the support of the Department of Energy, which is intended to overcome some of the methodological limitations that characterize most prior studies of possible "greenhouse" impacts on agriculture and other industries. One of these limitations stems from reliance on GCM scenarios. The recent EPA study used two GCM scenarios and came out with a number of projections about changing crop yields and water resources, energy requirements, and so on. Since the scenarios themselves are so uncertain, particularly with respect to regional detail, people who wish
to diminish the significance of such studies can wave them away by arguing, "Well, the scenario is unreliable." There is no perfect way to generate climate change scenarios. In our study we take a different approach; we analyze how a particular region of the United States -- in this case the four-state region of Missouri, Iowa, Nebraska, Kansas -- would respond to a replay of the 1930s exactly as it occurred.

I don't want to take up time with more details, but in order to overcome the reliance on GCMs, we use a real climate analog. In most impact studies the climate expected to occur 50 or 70 years from now is imposed on the world of today: it's the "dumb farmer scenario." The results assume a passive response -- no adaptation, no change in technology or management. However, we know full well that people are going to adapt. It's ridiculous to think that if climate is changing, people will not be searching for ways to adjust and adapt. So we're trying to establish a base description of this region and how it functions. We also will try at first to see what happens to the "dumb farmer." We will compare that model with one in which we put in the adaptations and adjustments that can be done with today's technology and today's knowledge base. That is the "smart farmer scenario." Finally, when this is accomplished, we're going to try to anticipate what the agriculture, forestry, water resources, and energy demands in this region will be about 20 to 40 years from now by which time, presumably, a significant climate change will have happened. Again, we will apply a serious climate change -- that of the 1930s decade. The 1930s were very severe. The GCMs are sufficiently credible in indicating warming and drying in this region so that we cannot ignore the possibility. We feel that if people can adapt to this worst case, there is some reason for optimism.

Finally, I would like to refer to some of our work which is reported in *Greenhouse Warming: Abatement and Adaptation* (Sedjo and Solomon 1989). The book is based on a symposium that RFF organized about 1-1/2 years ago. Many of us believe that, no matter what attempts are made to control greenhouse emissions, the reasonable and pragmatic thing to be doing now is to prepare for some (hopefully moderate) climate change. If, indeed, the climate does not change beyond its current normal oscillations and variability, we will, at least, have learned how to deal better with that variability. Droughts, floods, and cold spells are happening now. We don't have to invoke the greenhouse effect to justify concern with the sensitivity of the world food system to climatic stresses.

Sanchez: I don't think there's any question that plants respond to CO_2, and that they're more drought and cold tolerant; but that additional growth requires additional nutrients. In the liberally fertilized fertile soils of the United States, that may be sufficient to account for additional yield increases. But in natural systems there might be a limit to how much plants can respond to additional CO_2 if you don't have additional nitrogen and phosphorus and other mineral nutrients coming into the system.

Rosenberg: Let me respond in two ways. Growth chamber and greenhouse research with carbon dioxide fertilization have also shown that CO_2 moderates the effect of drought stress, salinity stress, and to some extent, phosphorus deficiency. The case for nitrogen is less clear. Under these stresses plants don't do so badly if they're fertilized with CO_2 as they do if they are not. So there is a kind of natural feedback by which elevated CO_2 concentrations might reduce the impact of certain environmental stresses.

Some time ago I wrote (Rosenberg 1982) that if the atmospheric CO_2 concentrations were to increase, there should be a rapid response in the rate of photosynthesis that would lead to greater biomass and, ultimately, to deeper and better root systems. Isn't it possible that with deeper, more vigorous root systems soil formation and mineralization would be accelerated? I speculated that a CO_2 increase, all other things equal, would have the effect of improving the condition of forests, especially by increasing soil organic matter content and water-holding capacity.

Davis: I don't think it's clear that would be the effect because the change in the C:N ratio in the litter might affect mineralization rates on the forest floor. But exactly how is unclear. Many forests are more strongly nutrient than moisture limited.

Ruttan: Do we know anything about the behavior and impact of soil micro-organisms? The literature seems to be rather empty.

Davis: I don't think anyone's done any experiments with enhanced CO_2 in an intact forest soil. The only comparable experiments I know of are in the tundra. There, the overall productivity has not gone up but the species composition has changed. In the forest you might expect species to respond differentially. But it's not clear what will happen.

Munson: Is there any long-term evidence that as you increase the organic matter in high fertility fields you enhance yields by raising the level of CO_2? I don't believe there is any solid evidence.

Abrahamson: I didn't understand what Rosenberg said about the carbon cycle. Are you suggesting that an increased amount of carbon has been sequestered in the biota? What's the evidence?

Rosenberg: None that is yet reliable. However, we know that half the carbon emitted by man's activity into the atmosphere does not remain there.

Abrahamson: Of course, it's going someplace. That's true. But the rate of carbon uptake and the amount of carbon sequestered are quite different things. I'm not aware of any evidence whatsoever that additional carbon has been sequestered in the biota or soil.

Rosenberg: I don't think there is any direct evidence. One of the most difficult things to establish is whether yields are increasing due to higher levels of CO_2, given all the natural variability in yields. I agree that I don't know of any evidence that this is happening. But I do know arithmetic. And it's either in the oceans or on the land. And the oceanographers have reduced their estimates of the oceanic uptake of carbon dioxide quite considerably from what it was 10 years ago.

Rawlins: That is a very important point. No comprehensive inventory of soil carbon storage exists, particularly carbon below the plow depth. We know, however, that even in forests half the carbon is in the soil. It's just not as visible and as easily counted as that in the trees. In one instance where deep cores were taken from rangeland on Blackland soil at Temple, Texas, 15 times as much carbon was stored in a hectare of soil as exists in the atmospheric column above it. Although not all soils are this high in carbon, these data illustrate that soil carbon storage can be large. A small percentage change in soil carbon storage would have a big impact on the atmospheric concentrations. We are aware that surface soil carbon is lost when soils are plowed. But what happens to deep soil carbon storage? If, in response to the 25 percent increase in atmospheric carbon dioxide that has occurred since pre-industrial times, plant roots grow deeper and are more prolific, isn't it possible that the soil carbon reservoir is being filled, not depleted?

Crosson: Is soil science sufficiently well developed so that you could theoretically say that it's possible for the soils to absorb these greater amounts of carbon?

Sanchez: I could try that. Stan Buol and I were asked to make an estimate of what would happen to soils, assuming a $3°$ temperature increase. First, the $3°$ increase is no big deal in terms of releasing additional nutrients. Second, there would be a net loss of carbon to the atmosphere because of additional temperature. But we estimated that the amount of additional litter needed to counteract that loss would be on the order of about 500 kilograms per hectare per year. The question is whether the CO_2 fertilization effect would be sufficient to counteract that.

Cheng: Let me respond to some of the soil carbon issues. A major deficiency when we looked at the biosphere is that we only considered the plant part and not the micro-organisms. That's one area where I've seldom seen any good data or even estimates. But what we do have certainly suggests that the micro-organisms may account for a large component of carbon, even when compared to plants. One thing we need to look at is the seasonal effect of the micro-organisms. It could be that they absorb more carbon because of summertime activity. So there could be seasonal variations. We need to look at the seasonal effect rather than at just constant accumulated effect.

Larson: The rule of thumb here in the Midwest is that we've lost one third to one half of our carbon in our cultivated soils. That may be a hundred tons of carbon per hectare -- a lot of carbon. Our studies pretty well agree that, under a given management system, the carbon level in the soil is in equilibrium with what is put back. If you plow back two tons per acre, you get a certain equilibrium level; if you plow back four tons, you get another level. Experiments that I did in Iowa 20 years ago clearly show that. With modern tillage practices and modern residue practices, there's a good chance that we can increase the carbon content of the soil. There's some evidence that that's happening. It is a research issue that we need to look at, not only what these carbon levels are, and try to develop management practices that will increase the carbon content.

Cheng: I want to raise another question. Rosenberg was talking about plant response to CO_2. The CO_2 fertilization effect is probably slowing the atmospheric accumulation of CO_2 below what it would be otherwise. From an agricultural production point of view, we know that the effect will be an increase in biomass. Will that necessarily translate into crop yield increase?

Rosenberg: Yes, I think so. Everything I've seen indicates that for the economic crops -- corn, sorghum, and millet, the C_4 plants -- there is an increase of 5-15 percent in harvestable product; in the case of C_3 plants -- soybeans, wheat, barley, potatoes, etc. -- the increases range from 30-80 percent with doubled CO_2 concentration.

Cheng: In yield?

Rosenberg: In yield! The only thing that would stop the biomass increase from turning into a yield increase would be a blocking of translocation, perhaps by starch accumulation -- i.e., the lack of a sink for the photosynthate. And there seems to be some evidence that this does not really happen. I saw a paper recently suggesting that kind of blocking mechanism is not interfering with the conversion of photosynthate into harvestable yield.

4

Global Climate Change and Agricultural Production

William C. Clark

Clark: Ruttan and I negotiated about two months ago that I would take a step back from the initial focus on global climate change and try to sketch a framework of resource and environment constraints that might impinge on agriculture production and agricultural policy over the long run. This may violate the first precept of agricultural policy analysis, which is don't go long and large: focus at the farm level. What I will try to do is to sketch a framework within which anything we say and most of what we don't say today and tomorrow might fit. It's put forward to provide some sort of counterforce to the pressure to focus too heavily on the very important issue of climate change. The issue is not so much whether immediate climate change effects are going to be large or small but, rather, to say large or small relative to what? That is, when and where might climate-mediated constraints come to the top of that list even though they may rank relatively low at other places and times.

My point of departure is a historical perspective: agricultural activities, broadly conceived, have been the primary transformer of the global environment for all of human civilization until very recently. One can still argue today that, at a global level and averaged over more than 10 or 12 years, it is still basic land use transformation activities that have been, if you had to rank them, the single largest transformer. Obviously, fossil fuel emissions, climate change issues, and ozone depletions give one pause. But if one tallies across lots of environments, the statement could be argued.

An equally defensible contention is that agriculture has been the sector primarily affected by environmental change around the world over the last several centuries. There may be impacts elsewhere -- on transport, on habitation, on human health -- but the place that so many of these changes come to roost first is, not unexpectedly, on human activities related to agriculture and natural resources. What that means is that the coupling of agriculture and environment, over large scales and the long-term, is a two-directional issue.

When we take up the notion of constraints around which these conversations are organized, it becomes important to look at agricultural activities as a source of many resource and environmental changes and, in turn, to look at the impact of these changes on agriculture. Agricultural activities can be affected in two ways that involve the environment and resource base. One is direct: climate changes and crops grow differently. The other is indirect as society becomes concerned with environmental degradation caused by agricultural or other natural resource exploitation activities, and adopts policies with multiple effects.

An old but useful example is the banning of DDT several decades ago. It was not that there were such bad impacts of DDT directly on the agricultural sector. Rather, society, at least in North America, decided that other environmental consequences of the use of this chemical affected agriculture and health. Decisions were made that then had potentially severe impacts on the agricultural sector. I think we've got to be careful as we consider this second class of constraints when policies initiated to protect or regenerate a component of the environment have immediate negative impacts on agricultural production.

As I tried to figure out what Ruttan was doing with this workshop, it seemed to me that it is the notion of constraints -- translated into an effort to understand when and where we might expect environmental constraints to become important enough to change agricultural production or policies affecting agriculture.

Somewhere, one has to think about taxonomy.

First, we have to end up, implicitly at least, with a list of which constraints belong on the list. What kind of constraints should we be check-listing to see whether they will end up having a big or little impact on agricultural production and policies in given places?

Second, we have to identify the kinds of human activities and natural processes that might change the nature of these constraints -- increasing, decreasing, or mitigating them.

Third, we also need to say something about the rates of change of those constraints and the activities that generate or remove them. We need to get a feeling for which of them are changing very quickly and which, very slowly. Those that are changing relatively rapidly may be expected to pose greater challenges to production and policy than those that are moving slowly enough so that we can adapt our practices and policies as we go.

Finally, we have to ask what these possible constraints mean in particular agricultural production settings: to particular farmers, researchers, extension agents, and national policy bodies. An effort should be made to do some first-order guessing on which kinds of places on earth might be expected to incur the same kinds of constraints at the same time.

Let me say a couple words on each category. I'll say very little about what's forcing them. Rather, I will talk about how we might try to think about the rates of change in those constraints and sorting them out in different parts of the world. As part of some other work that I and others are doing, I have come out with a list (anyone is free to steal, savage, or amend it) of eight different categories of broad constraints on agricultural production and policy in the large and over fairly long periods. They are grouped by order but are not priority ranked. In all the categories -- which will be obvious things like land, water, and energy -- it is useful to talk separately about the quality and quantity aspects.

Land clearly can be a constraint. The area available for agriculture is limited. This is reflected in the arguments about slowing deforestation and in the notions that habitation is encroaching on the most productive land. The area available is the quantity dimension. The quality dimension is where all the issues of soil productivity or fertility come in and the various ways in which the location properties of land can be assessed: the erosion, the salinization, acidification, compaction, and another class of things we talk about when we talk about soil degradation.

The second, obviously, is *water.* In quantity terms: how much can we get? The answer must be viewed in terms of some sort of supply function, that is, how much could we get if we were willing to pay a certain something for it, broadly conceived in terms of trade-offs, against other things society wants? Simply put, the quantity competition is among irrigation water, groundwater, and other uses. Independent of the issue of climate change, the constraints on quantities of water for agriculture is likely to become dominant in many parts of the world at certain periods over the next century. (Not all parts of the world and not all periods.) A quality issue also is involved here. There are places where the quantity of water available is perfectly sufficient but the quality, in terms of salinity, the toxicants it carries, or, in some cases, the human pathogens that live in it, represents an actual constraint on agricultural production.

Third on my list is *energy*. Quantity issues include how much is going to be available to the agricultural sector and at what prices. The quality issues include what kinds of energy will be available: low head hydroelectric, liquid fuels, high grade or low grade solid fuels, or others. For some regions, some of these issues seem to sit very low on the list of present or likely future constraints. For others, dominant impediments to doing certain kinds of things in agricultural production that you would like to be able to do but can't are quality or quantity shortfalls in energy.

The fourth is *fertilizers* or generic nutrient additives. The quantity issue is, again, how much can you get at what price? A quality issue, one we usually don't tend to think of enough, is what fraction of it is lost? What fraction ends up as a potential pollutant rather than an incorporated nutrient in the plant? This is far enough from my own area of work that I don't even know the proper literature. But a lot of work is being done on capturable forms of nutrient additives versus forms that, when you plop them down, most goes into the air or into the groundwater. There also are some other quality issues. The one Europe may be facing now is the trace metal pollutants in fertilizers that end up as a residual in the soil slowly or are incorporated in plant tissue. That's surely a quality issue which we need to look at over the long run. One European country recently decided that within this century it will be facing a situation in which it must stop eating its own foods because their heavy metal content exceeds safety standards. Alternatively, the country can revise its environmental regulations to raise the limit on heavy metals. Either way will be desperate news for the policy-making apparatus of the country. When you once imposed X parts per trillion of cadmium as a health limit, it's very difficult to come back later and say, "We fooled you; that was an arbitrary limit: it should have been twice as high. Now we can go on with business as usual." That is not a popular way to win an election.

Fifth is the *biocides* issue. Again, the quantity dimension is how much can we get at what price? Obviously, use can be constrained, as in the DDT case, for reasons other than direct negative effects on agriculture. On the quality side, we should be thinking about such things as, to what extent is this biocide something that kills absolutely only the beast you want to kill versus a broad-based one? To what extent is it one to which pests have built up a lot of resistance and therefore has lots of spill-over effects? Again, such issues, talked about specifically in policies about pesticides, don't tend to come into the broader discussions of the sort we're dealing with here. They probably need to be.

Climate is sixth. The quantity dimension must be things like temperature, precipitation, and evapotranspiration. The quality, and this is where the categories begin to be strained, includes seasonality: the difference, in a sense, between X centimeters of rain distributed evenly over the year and its falling in utterly useless and, indeed, counterproductive spurts. This, as you've heard this morning, is another concern that comes up in the issue of climate change.

The seventh is *biodiversity*. A lot of special pleading in this area has not been particularly useful. Biodiversity has been reaffirmed as a good in itself; it is a cornucopia from which we can't afford to spend one unit. If the quantity issue is the number of species, the quality issues are things like the supply of natural predators on the pests we're after. Do we still have them or have we lost them? What about future cultivars: the genetic base from which the new crops of the future will come? There is the old argument about possible pharmaceutical supplies. But there is also a quality issue in biodiversity. One challenge is to move from the broad brush prescription to never deplete biodiversity to a quantifiable notion of which aspect, which kinds of biodiversity, are most important. This list will contain certain sorts of predatory insects and certain first, second, or third cousins of things we know are cultivatable. We should begin focusing very tightly on those because we're going to be no more successful in preserving biodiversity for itself than we are in preserving land or anything else for itself. We've got to be more specific, and the specificity must come from the agricultural production sciences rather than the other groups that are flailing around in this area.

Finally, though it bewilders me, *agricultural output* often gets left off lists like mine. On the quantity side is the same old issue of how much can you supply at what costs, given all the other constraints society has loaded onto the system. The quality issues include the obvious ones that all agricultural production people worry about: what is the value added in a market sense by its attractiveness, its taste, and its nutritional value? We may also want to be a little more concerned with the portion of the crop, and its basic carbon and energy values, that are wasted, both in the field, and in the process of moving it to the table. We may also want to look at toxicity. More and more we are going to be faced with notions of different qualities of agricultural produce simply in the sense of whether a case can be made that it's pure: whether it's natural, a popular notion right now, or, in fact, loaded with at least some noteworthy quantities of trace metals, organic compounds, biocide residues, and the like. These issues are talked about in very small circles. They tend not to be raised in general discussions like ours.

That's the list. Obviously, subdivisions are possible within each item. But it would be a mistake, at least for the larger Rockefeller consultation exercise, if we were to become preoccupied with any one area, even when there's a window of opportunity to move rapidly into areas like climate change. Rather, we need to look at something like my list to specify for particular regions and times the changing hierarchy of constraints.

As one moves to the notion of what belongs high on a list of potential constraints, at least until you find a reason to drop it, I would emphasize things like energy use, fertilizer use, and biocides use. They are growing at extraordinarily high rates relative to anything we've had, say, over a 30-, 40-, or 50-year period in the past, and are plausibly accelerating. Changes in things like land area and quality and even water availability are happening at significantly slower rates.

Finally, I argue that some sort of a taxonomic effort might provide us with a smallish number of groups in which we might expect, should we ever begin to assemble the data, that the rankings would tend to be more similar within than between groups. One such taxonomy comes from some of Crosson's work which I find useful: the very simple notion of taking a two-dimensional matrix in which population density is plotted on one axis and wealth per capita or square kilometer or level of technological development is plotted on the other. With a third axis, which we both agreed we wouldn't try to do, you would probably want something about the nature of the landscape. Agroclimatic zones as indicators of basic biological productivity potential actually works pretty well as a sorter. But even just taking the four grids -- high and low on the population density and wealth density -- tends to produce a nice basis for ordering different countries or regions.

The research my students and I are beginning now, looking forward to some plausible trajectories into the future, suggests that those categories hold up well under various versions of the scenarios that might develop. Even when they break down we're finding it a better guide than nothing in efforts to look ahead to the kinds of policy problems that may be confronting clusters of regions 20 or 30 years into the future.

Rayner: Clark's list is interesting. Taxonomies are a useful way to begin looking at problems, but concern with the lists he's given us is that his categories seem to be very much dominated by a kind of supply-side view of the agricultural system. Do you not think it would be important in looking at constraints on production to consider not just the numbers of people but their age ranges and the kind of different nutritional demands that they're going to have also?

Second, given what we know from the intervention in markets in terms of subsidies and taxes -- including price support systems, forced delivery, and biased exchange rates -- the actual structure of markets, transportation, and trading arrangements also represent very real constraints upon agricultural output. To supplement his list, we need to look more carefully at the demand side of the agricultural system.

Clark: I couldn't agree more that somebody should do that. I took Ruttan's instructions to address the resource and environmental constraints on agriculture in this session, and to ignore things being addressed in other sections. I knew Rayner would be here so that there wasn't a chance that I wouldn't be reminded properly and articulately that too often sessions like this have started and ended by saying, "Well, of course we know that the social, institutional, and human dimensions are important, but for a moment let's just concentrate on the technical resource ones," and so on.

Crosson: If I heard Clark correctly, he is saying that there are things happening on the demand side that have an important bearing on the emergence of the various environmental constraints on agricultural production. If we believe that global demand for food, for example, will grow by 1-2 percent a year over the next 50 to 70 years, the implications for emerging constraints in agriculture are vastly different from demand that grows by 2-4 percent.

Ruttan: If population growth rates in Africa south of the Sahara were closer to one percent than 4 percent, it would have a dramatic impact on our perception of how to achieve sustainable growth in agricultural production.

Clark: In my notes for this meeting I had two columns but I only briefly alluded to the second. The first was the resource/environmental constraints: the list of eight that I just read. The second was the determinants of changes in those constraints. That is exactly where I put in the notions about whether you need a lot or a little; ability to pay a lot or a little; the structure of the labor force; and these sorts of issues. So I see it at least as a compatible direction for the discussion to go.

Munson: I presume, when you're talking about nutrients and the need, you're talking about worldwide, and you said they are increasing. The reason I raise the point is that in the United States, for example, phosphate use actually peaked in 1977. Nitrogen and potash use peaked in 1981 and have been declining since. I suspect the same holds for the pesticide total usage.

Clark: It may well be that the trends you note will hold both in this country and elsewhere. A couple of years back effort was made by the natural sciences community to try to say something usable about what kinds of chemical inputs into the environment might be occurring over the next 100 years. We got the scenarios worked out pretty well for things like greenhouse emissions, but very poorly for things like pesticide use or nutrient releases.

We went through an exercise trying to say for the various scenarios that are out there from official and unofficial organizations, what can they tell us about the likely rates of change? For whatever it's worth, when that stuff was worked through, the 1975 to 2075 vision for North American fertilizer use implied a 1-1/2 fold increase with most of the increase occurring over the next decade or two. This estimate is a lot smaller than the increases of four and five and eight times that are foreseen for other continents.

Ruttan: A friend of mine recently did a comparison for Minnesota and Saxony (Germany). Fertilizer inputs per hectare in Saxony are about four times higher than in Minnesota. Given the changes in population densities around the world agriculture, in terms of input use, will look more like that in Saxony, Japan, or the Netherlands than in Minnesota 50 years from now.

Clark: When we completed our exercise, even though the numbers are poorly grounded, nobody could erect for us a continent-scale scenario for fertilizer 100 years from now that brought levels of fertilizer application on a kilograms per hectare basis up to present European lands. This is quite interesting given the problems Europe is confronting today with its groundwater and its heavy metal toxin build-ups. It is important that we discover where they come from. One can subtract for fossil fuel, industrial material, and come out with a residual. The point is that such a study hasn't been done. It's hard to find anyone who really wants to do it. It is something that isn't being written up in the literature. This is a kind of report you get from skulking around the health bureaus and asking lots of questions. It's the kind of study that just seems to me to be enormously important.

Larson: Well, the cadmium story has been overdone. But I'm not saying that it isn't a problem in Denmark or in Europe. It is! But it's probably been overdone.

Rawlins: Are they using sewage sludge?

Clark: Yes, of course, in Denmark.

Sanchez: That's a source of cadmium.

Clark: But nobody has sorted out the relative importance of the several sources. I raise it only to say that, whatever its scientific foundations, it is typical of the kind of issue that is going to jump right out of the science arena and into the policy arena. It will then be thrown back to the scientists as a mess that has to be disentangled, probably with an inadequate data base and too much pressure for instant results. Given the predictability that it will become an issue, one might want to put it in place as a relatively low key research program that would provide the data needed to tell the story right, rather than merely responding because the Green party this year says that the phosphate merchants are poisoning us.

Sonka: I don't know if labor should be considered a resource constraint in this context, but in the United States and the other developed countries the 1990s will be a period of labor scarcity.

Clark: There is a strong historical precedent in this argument because one constraint we will face is not only the numbers, but the quality issue. This includes education and training, health, and others. They are obviously constraints in some parts of the world. If I'd been smarter, they would have been in there the first time.

Sanchez: As population increases more, more of the world's agriculture is going to look like that in Japan, the Netherlands, and

Saxony. What about places where population is already extremely high, such as China and Bangladesh? Do we expect them to go to an even more intensive, high-input system of agriculture?

Clark: This is one of the big questions. We have examples of countries that, starting out under high population density and very low input intensity, have made a transition to what may be ecologically sustainable forms of agriculture. They have done it by going the intensification route. I have tried to work through, as an intellectual exercise, routes of development -- especially rural development -- for our high population density, low income density areas that don't involve radical intensification of input use and increased value-added capacity of agricultural prediction in the rural areas but, for the life of me, I can't see one that works. Even solution of the food problem may not solve the labor absorption and income generation problems. When one thinks about the environmental problems associated with agriculture that are likely to confront such areas over the next 20, 30, 40 years, you just have to work within the constraint of finding alternative ways of producing a lot more value on the same land. Hopefully, these economies will develop in a manner that will enable them to get some people off the land so that they don't get caught in this horrible partitioning of units of production down below a usable scale. But that's tricky. When you then start looking at the levels of inputs that would be required under known systems to produce the levels of output or value added we're talking about, they include extraordinarily high densities of fertilization, biocides, and, in some cases, energy application that will have some very serious regional-scale, environmental implications.

Sanchez: So alternative agriculture, as we understand it in this country, doesn't seem to have much of a future in your scenarios.

Clark: Well, you know, so much comes under alternative agriculture. But I don't see it solving the income-generation problem for enough people.

Ruttan: I have two questions that I'd like to get a response to. The first concerns the investments that would be required either to reduce the sources of global temperature change or to respond to temperature change. We have now brought up the issue of the kinds of investments it would take just to sustain present per capita income levels, or to raise them at some acceptable rate, in most of the world. If we honestly attempt to face these issues we must be talking about levels of saving and investment that our societies are not yet prepared to achieve.

The second question may not appear sensible to people on the physical science side, but I've often wondered -- economists often think in terms of optimum -- is there an optimum level of CO_2 and associated greenhouse gases? It strikes me that we often start out by asserting that there was a golden age, between 1850 and 1900, when the CO_2 level was about right. If we had been doing an experiment, we would have designed it to answer the question of whether we would be better off by reducing or enhancing the CO_2 level.

Jones: To answer that question you have to ask, "better for whom or for what?" From a biologist's standpoint you assume that all animals and plants have evolved with a certain level of CO_2. That suggests that the optimum is what it is at a given time in evolution.

Ruttan: But apparently it came down from some higher level in the distant past.

Abrahamson: It was higher.

Rosenberg: Let me take a crack at the second point. In an adjoining state you have a well-known climatologist, Reed Bryson, who, 10 or 15 years ago, was telling us that we were heading into the next ice age (Bryson, 1974). I don't think there's anyone who can dispute that argument because we know we're in an interglacial period right now. When the fact that the climate was not getting colder was thrown at Bryson, he answered (in essence), "Well, it's probably the CO_2 that's masking the cooling effect." Whether he's right or wrong doesn't matter at this point but, clearly, climate does change. Suppose we decide that we like the climate the way it is and suppose we're farsighted enough and our models are good enough to show us how to stabilize climate the way we like it. Then we might even be advised to pump still more carbon dioxide into the atmosphere to give us a little extra greenhouse effect. Unfortunately, we don't know just how to fine-tune the system. If we were confident that more CO_2 would not produce significant climate change, I would say let's pump the stuff into the atmosphere. I can't see where it would do us any great harm. We are existing in a room now with a CO_2 level considerably above the ambient, and I don't see that it's hurting anybody.

Just to finish this thought: the concentration today of 350 parts per million is not threatening and does not appear to be threatening. It's conceivable that we could go to 400 or 450 parts per million without running into any real threat in terms of the biology or the agricultural effects. However, if the concentrations rose toward 600-800 parts per million, then there could, indeed, be certain species or subspecies that would experience significant negative effects.

Abrahamson: The first question is about the increasing demands for investment and savings. I would guess that we're talking about diverting about the same amount of resources that we're now spending on defense. To the optimal level of the CO_2 question, I would respond by stressing the rate issues. If rates of change are such that significant climatic change occurs in a time less than the lifetime of people or trees, there will be trouble. I don't think the question of optimum is interesting. Much more interesting is what rates are tolerable or what costs are associated with adapting to or mitigating the various rates.

Chen: In response to that, it depends on whether you're moving to something good or bad. If you're moving quickly to a better world, some people may be harmed, but perhaps there will be fewer objections. It does matter whether the transition involves a positive or negative change. I don't think it's valid to say you should only look at rates.

Abrahamson: I don't say only look at rates. There may be other things, like threshold phenomena, that you need to be concerned with. I'm worried that you're beginning to sound a little bit like Budyko (1989), the Russian climatologist, who argues that the next century or so is going to be pretty bad. But in another 100 years, if we keep on growing fast and if we keep on warming fast enough, things are going to get better. We should just accept the transition and get through it as quickly as we can. That argument is just silly!

U.S.D.A. and Climate Change

Rawlins: We've already discussed a number of issues that I think are important. I'd like to add three points. First of all, Clark's comment is important. Not only is agriculture responsible for some of the constituents that are capable of changing global environment, but it could be seriously affected by these changes also. Uncertainty about future environmental factors that could affect food security is frequently used as a primary justification for the U.S. global change research program. But the resources available for direct study of food security issues are small. In the past, USDA has not assumed primary responsibility for assessing either agriculture's contributions to global environmental change or the effects of these changes on agriculture. But the level of interest is changing.

I returned to the ARS National Program Staff a year ago at a time of awakening to global change issues. Dr. Orville Bentley (Assistant Secretary of Agriculture for Science and Education) had just learned that USDA was not represented at the organization meeting of IPCC at Geneva, and he wanted to know why. He informed the Department of State in no uncertain terms that USDA would be represented in the future, and he appointed some staff members as representatives. Up until then, Dr. Norton Strommen, Chief Meteorologist with the World Agricultural Outlook Board, an agency serving under the Assistant Secretary for Economics, was being called upon to represent USDA on nearly all issues related to agricultural weather and climate. The midwest was just emerging from a devastating drought, and Strommen could not cover everything. Dr. Gary Evans, ARS Deputy Administrator, was given the responsibility for developing a USDA Strategic Plan for Global Change Research and for coordinating USDA representation on national and international planning groups. As a member of Evan's staff, I was asked to chair the plan development process. Until then, without the needed people to cover all the bases, USDA had taken a back seat, basically just monitoring what was going on in CES and other groups. Since then, Dr. Evans has been given the assignment as Special Assistant to the Assistant Secretary for Global Change Issues; he chairs a very active working group from all USDA agencies to make certain that USDA carries its share of responsibility for global change research.

Allen: I'd like to know who came to the conclusion that agriculture shouldn't be involved.

Rawlins: I don't think anyone really made a decision that USDA should not be involved, I think it was just a matter of too many issues and too few people. Norton Strommen had traditionally represented the Department on agricultural weather and climate issues, and when new issues related to climate came over the Secretary's desk they were routed his way. He chaired the USDA Climate Coordinating Committee, which includes Science and Education Agencies, but this group was insufficient to handle all the rapidly emerging issues.

Ruttan: Has the Department yet developed an agenda for research that will enable it to deal with this nexus between climate change and agriculture?

Rawlins: Yes, the USDA Strategic Plan is nearly complete and makes a good start in this direction. But we have a long way to go. It's my opinion that we have some substantial barriers to overcome.

The second point I'd like to make is this: Although USDA at one time was the elite research agency in government, research is now only a small part of the USDA agenda; it represents something on the order of two percent of the budget. We have to be realistic. If you were the CEO of an enterprise with a division that represented only two percent of the action, how much attention would it get from you? I think it's going to be very difficult to generate the budget required from within USDA to adequately address the global change issues without some strong support from outside. Each year every agency within USDA competes for budget. For the last few years the pressure has been strong to keep the departmental budget constant. It will be very difficult for a small program to obtain a substantial increase if it has to come at the expense of other programs within the Department. Our best hope is to build strong cooperative linkages with other members of the science community and to obtain their support for our budget. Perhaps this issue should be addressed as an institutional problem in a subsequent dialogue.

Rosenberg: In fairness, one should point out that a lot of the research over the last 10 or 15 years on the direct carbon dioxide effects on plants has been done by USDA. But it has been done with DOE money. This tells you something about the level of importance attributed to this work by the USDA establishment.

Rawlins: This is a problem. USDA scientists are sought out by others because of their scientific credentials for dealing with the environmental issues we're facing. USDA has not been successful in obtaining funds to pursue these issues directly. Other agencies are successful in obtaining the funds, but they often lack the scientific expertise needed to solve the problems. USDA is not viewed by everyone as being unbiased in addressing environmental issues. Some perceive agriculture as being part of the problem, not part of the solution. That's an image we need to do something about. I don't think agricultural scientists deserve this image but, again, we are only a small part of a very large Department that has constituencies who take adversarial positions on environmental issues. The constituency for agricultural science must be extended beyond people who live on farms. A substantial part of the scientific expertise capable of dealing with environmental issues in the managed ecosystem are within the agricultural science community.

Ruttan: I have a couple questions. I'd like to direct one to Allen because he has been sitting on the NAS/NRC Board on Agriculture. The Board came out with this dramatic proposal for a $500 million increase for agricultural research: a relatively large increase. Did the set of issues we are discussing enter into that proposal?

Allen: It is one of the six categories that has been targeted for increased support. It's even broader because many things that are targeted, such as plant systems, are related to these issues.

Rawlins: The Board on Agriculture report has been helpful. It has helped to gain attention and visibility for agricultural research. The third point I'd like to make is that food security is the actual issue we should be concerned with. It's related to the question of who is our constituency. We need to ask ourselves some tough questions. What is our central concern? Is it food or is it farmers? If it is food, then we may design a very different production system than if it is farmers. When we talk about environmental and technical constraints on sustainable growth in agricultural production we need to know what kind of an agricultural system we are talking about. How far are we willing to allow ourselves to depart from the traditional system, with food production being carried out by farmers? Are we willing to consider alternative systems that might completely redefine the role of farmers?

Martin Rogoff and I (see Rogoff & Rawlins, 1987) tried to take a look at the fundamental constraints limiting food availability in the future. We argued that one of the biggest limitations is that the productive capacity of our present system varies dramatically from year to year in response to weather. Our present system is based primarily on the seeds of annual crops. Storage of the perishable products from farms is expensive, so the carrying capacity of the system tends to be controlled more by years of minimum production than by average production. It's pretty much a hand-to-mouth system. If weather fluctuates even more in the future, the carrying capacity of our present food system could decrease. Furthermore, annual crops are not well adapted to most ecosystems so fertilizers, pesticides, and irrigation are required to create a hospitable environment.

Ruttan: Nature abhors agriculture!

Rawlins: I agree! If we look down the road 100 years or more we see that the inputs needed to provide the environment for annual crops will become limiting. The stark fact is we do not have a food system that can outlive our fossil fuel and fertilizer supplies. Farmers in developed countries use more calories in the form of inputs from fossil reserves than they produce in the form of food.

As an alternative to producing the whole crop in the field, we considered the possibility of using the field only to capture the sunlight energy and store it as carbohydrates or lignocellulose in perennial crops, including trees. These products would be harvested as needed to produce sugar syrup, from which food products would be produced biotechnologically off the farm. The raw food products would be stored as a living reserve of standing biomass. The plants could be chosen on the basis of their adaptation to the ecosystem to reduce inputs. Since nearly all the energy captured could be converted to food, rather than just the seeds, the capacity of the system could be substantially increased over existing systems. A first step would be to convert lignocellulose to animal feed. This alone would release a large amount of land for other purposes. Other products for direct human consumption would follow as the market demanded.

The point is that in considering environmental and technical constraints to agricultural production in the future we need to consider possible designs for new production systems as well as resource limitations for our current production system. Certainly we would not want to be accused of considering only the dumb food producers' scenario.

Ruttan: You've outlined an alternative agriculture that the adherents or promoters of "alternative agriculture" would not embrace.

Rawlins: You are probably right.

Munson: Rawlins referred to the amounts of energy that are used in agriculture. In the concern with CO_2, energy is the overriding issue. If we're actually interested in reducing CO_2 emissions, we should spend money on nuclear fusion research rather than on putting a space station up and going to Mars. They're talking about $400 billion to be spent in the 1990s on that effort.

We put half of our corn crop back into the soil. There's a harvest index of roughly 50 percent. So for every bushel you take off there's an equivalent amount of dry matter produced that's going back into the system that could be harvested and used.

Ruttan: When Rogoff first mentioned to me over the phone the work he and Rawlins were doing on biomass, my first reaction was, "That's great. We're going to move agricultural production from the temperate regions to the tropics." His reaction was, "Don't tell anybody at USDA!"

Chen: Rawlins, you mentioned that USDA is getting into the global change program. I would assume that USDA's interest is primarily on the impacts and policies side? Is USDA going to actually jump in with more research?

Rawlins: Our main new thrust is research in support of the CES Global Change Research Program (CES is the Committee on Earth and Environmental Sciences of the Executive Office of Science and Technology Policy's Federal Coordinating Council for Science, Engineering and Technology). We will also be involved in impacts assessment and development of response and policy strategies.

Chen: That's my question. I have watched these things for 10 years. The research on inputs and climate tend to suck up all the money. There's a lot of lip service to the impacts and policy but in the end no one actually does any work.

Bochniarz: I'd like to get into another question. Fusion is an interesting prospect. But what about other alternative sources of energy? It is possible that if we had kept expenditures for solar energy at the level of the middle 1970s we would have, in the middle of the 1980s, economically viable sources of solar energy. All big companies have completely cut out the research on solar energy.

Rayner: I'm sorry, but at least as far as the United States is concerned, the question of alternative energy technologies isn't anything near as interesting as the issue of energy conservation through efficiency. Current actual U.S. emissions averaged about 1-1/2 gigatons for 1988.

The Edmunds-Reilly base case projection to the year 2020 suggests an increase to something over two gigatons of CO_2 emissions for 2020. The best saving that we can predict at the moment with nuclear, solar, and biomass energy, would still bring us down to something just over 1-1/2 gigatons of carbon emissions for 2020. On the other hand, energy efficiency improvements alone would bring about an actual cut from our present emissions level by about one-third, and to a halving of the predicted emissions level from the Edmunds Reilly base case. If we then add on the alternative technologies for the year 2020, we can cut down to about one third of the present emissions levels. It's not until you get into the 50-year time scenario that the availability of improved or non-emitting energy technology has an important impact on U.S. fossil fuel CO_2 emissions. In the United States, as far as getting an immediate big bang for the buck is concerned, we should be talking about efficiency not about alternative energy technologies. That situation is quite different for the developing countries.

Allen: Please give me some examples of the high efficiency scenario.

Rayner: Examples include increased efficiency in electric motors, in commercial and domestic applications, improved generating capacity, and improved transmission.

Waggoner: Are all the assumed increases in efficiency within the range of present knowledge or technology that is on the shelf?

Rayner: Absolutely. Those are things we can technically do tomorrow.

Crosson: Who is paying attention to the economics of conservation?

Rayner: The economics are a different problem and the situation in the developing countries is different from that in the United States. We expect the developing country emissions to exceed those of the United States by about 2000, and we're expecting that on the assumption that developing countries will use a lot of fossil fuels. This is what I wanted to raise with Rawlins. He talked about moving to a perennial cropping system based on lignocellulose for food security. One of the things we've been considering at Oak Ridge is that if you stop developing countries from moving into fossil fuels, what kind of alternative generation technologies can be provided for them? We will not realize the same efficiency benefits in developing countries as in the developed world because the benefits rely on an inelasticity of demand for energy services. In developing countries energy demand is very elastic, particularly in China. We've been very interested in looking at the whole issue of biomass for energy in which the biomass is gasified and the gas is run through a high efficiency turbine. Bob Williams of Princeton has made some estimates which show, in fact, that it is a cost effective technology. In relation to CO_2 emissions, it has the benefit of closing the fuel cycle because you are fixing the same amount of carbon in fuel as you're producing. Presumably, what we would do is woody biomass; things like fast-growing sycamores on plantations. It wouldn't appeal to the environmentalists from an aesthetic perspective -- you wouldn't have nice parks and forest out there to feed this energy system. And I suspect that if you were using a similar technology for your lignocellulose production, it also would not be the parks that would be producing the biomass.

The question is, what is the capacity of the agricultural system really to make a transition toward growing these kinds of woody biomass crops? Also, what would be the social acceptance of such an approach?

Allen: I can add a supportive example. We estimate at the University of Minnesota that we could decrease energy use by about 30 percent with some very simple changes in high efficiency light bulbs, improved efficiency air conditioners, and other equipment.

Rawlins: I suspect that the total annual biomass production rate of perennial plants is about the same as good annual crops. But our scenario would convert a larger fraction of it to food than our present system does. The reason biomass energy production requires high biomass production rates per unit area is the hauling cost for the raw input product. This, in turn, is related to the economy of scale of the conversion plant. Energy conversion plants must be large. I don't know how large the plants would have to be to convert lignocellulose to sugar syrup but, hopefully, they could be smaller or even portable so that more extensively grown biomass could be harvested.

Abrahamson: You said something to the effect that the economics are another matter, and I don't want to leave that. I've been involved for 15 years, I guess, with these energy analyses, particularly in Sweden, and it's clear that even though the unit price of energy may go up, the cost for energy services can go down. Reduction of emissions in the range you suggested can be accompanied by net savings in total cost of delivered energy service.

Rayner: Absolutely. And you could even reduce the price of the electricity as well. But the economics are more in terms of the rate of capital turnover and not so much in the cost per kilowatt hour delivered.

Abrahamson: Would the user pay less for energy services under the low-emissions scenario?

Rayner: Yes.

Sanchez: This lignocellulose scenario is very exciting to me because I work in the tropics. Obviously the place to produce it is in the tropics.

Rayner: And labor is relatively inexpensive there.

Sanchez: In the tropics we can grow biomass in 2-3 years that would take 40 years to grow in Minnesota. There are clearly some geopolitical implications here. But I want to make an additional point. Oil palm can produce about five tons of oil per hectare per year. During the Second World War, the French, using a simple filter, adopted oil palm as fuel in their diesel engines. However, oil palm use for food is going to decrease because of the cholesterol issue. We should look at these and similar tropical plants to produce oil directly.

5

Forest Response to Climate Changes

Margaret Bryan Davis

Ruttan: In her research Davis has been looking backwards over very long periods, whereas we take 50 or 100 years as the long term in most of our projections.

Davis: It might be a function of the lifespan of the organisms I'm working with. The long life span of trees makes their response to climate very different from agricultural crops. Let me summarize what I think will be the response of forest to climate changes in the future.

We have calculated the shifts in geographical range that you might expect to find in the northern hardwood forests of the northern United States, given two different climate scenarios projected by different General Circulation Models (GCMs) for a doubling of atmospheric greenhouse gases ($2 \times CO_2$). The shifts of geographical range for the trees we worked with -- sugar maple, yellow birch, hemlock, and beech -- are very large. The northern limit would move northward between 500 and 1000 kilometers, and the western range limit would retreat eastward. If you use GCM output that projects both rising temperature and an increase in rainfall, the range shift is about 500 km northward. There is a slight retreat from the western range limit, because the temperature increases are large enough to create moisture stress for trees. With a GCM scenario that predicts a large temperature increase and a large deficit of soil moisture in the central part of the continent. The range shift is much more extreme: the new potential range hardly overlaps with the old range except far to the east in northern New England and Nova Scotia.

The conclusion to be drawn from this exercise is that changes in forests will be very near species' range limits. What happens in the center of the species' range depends on the degree of ecotypic specialization in the species. The question is, can a beech tree that grows now in Nova Scotia tolerate the climate of Georgia where beech trees grow today? Or are the beech trees that grow in Nova Scotia specifically adapted to the Nova Scotia climate?

This kind of information is in short supply for many tree species. More is known about trees that are planted in commercial plantations: some species have a broad tolerance range and others do not. But for trees that grow in unmanaged forests, even valuable trees such as sugar maple, which grows so abundantly that nobody ever bothers to plant it, there has been very little research on which to base relevant models.

The long life span of trees creates problems for forestry that differ from agriculture. A forester needs a precise prediction of what's going to happen 50 years from now because the turnover time for many forest plantations is about 50 years. Some trees can be cut after 40 years, but the rotation time for others, such as oak, is in the 90-year range. Who can tell a forester the appropriate tree to plant, given that it will not be harvested for 50 to 90 years?

Pessimism has been expressed about the accuracy of the GCMs. It is my impression that most GCM modelers are trying to give us a general idea about the trajectory that climate might take. They might be rather alarmed to see me taking their data so literally and trying to project a range shift. We need to keep in mind, as well, that doubling of CO_2 is not an equilibrium condition, nor even an upper limit. Greenhouse gases will continue to increase until we run out of fossil fuel. Temperature thresholds will be reached eventually, although the predicted timing varies with the particular model. For these reasons, it is more useful to think in terms of rates of change than to project equilibrium conditions with doubled CO_2. Certainly for trees it is more useful, because the effects on trees are different with a slow change than with a fast change. If there is a slow rate of change, then the trees themselves will not experience a climate change during their lifetimes that is large enough to actually kill them in situ.

Historical experience tells us what to expect. During the droughts of the 1930s, hemlock trees growing near their southwestern range limit in Wisconsin died from drought stress. In the first year of drought they lost more than 80 percent of their root capacity, and in the second year, they died from insect damage. It is surprising, however, how little hard information there is in the literature about direct physiological stress to trees caused by climatic events. Information on growth response is available from tree-ring studies, but there are few documented cases of trees being killed by a climate extreme. This is why we know so little about climatic thresholds for adult trees. Much more information is available about the sensitivity of seedling stages, although even here the information is sketchy for most species. Flowering, fruiting, and seed germination and establishment appear to be most sensitive to climate. For this reason the first effect one would see in a forest would be the failure of reproduction of canopy species and the invasion of the stand by new species from outside the forest. We have seen major turnovers of species composition of forests over the last century because of our logging activities. In this part of the country, most of the landscape supports early successional tree species, such as birch and aspen. To have another major turnover in species composition of the forest might not be more surprising. I think humans would cope with it.

If, however, the climate changes very rapidly, then the effects on forests will be different. Trees will experience problems in dispersing into new areas where they can grow. The distances are much larger than trees disperse normally over the course of a few decades or even several centuries. Furthermore, the climatic changes could be so large during the lifetime of a single tree that from the standpoint of a forest manager, it becomes a question of whether there is any tree species for which the climate is suitable both at the time the tree is planted and at the time the tree matures and is ready for harvest. This would be true even at rates of change less extreme than those quoted by Abrahamson. A climate change of say, 5°C over the course of 200 or 300 years would exceed the tolerance of an individual tree.

Let me turn to the direct effects of CO_2. It is difficult even to speculate about the direct effects of CO_2 on forest ecosystems, considering soil as well as trees. It is much harder to do field experiments with such a complex system. The fossil record indicates that trees lag behind other organisms in their response to climatic change. The lags were due either to limited dispersal of seeds or to slow development of suitable soils. This is so, even though the climatic changes in the past were at least an order of magnitude slower than the changes we are projecting for the coming century. The generalizations I am making about response to a rapid or slow climatic change, which are based on a conceptual model of what might happen, are supported by results in the fossil record, which show trees lagging for decades or even centuries in their response to climate, relative to other components of the ecological system.

Waggoner: Does that mean the tree survives into the unsatisfactory climate?

Davis: Most rapid climatic changes that we know in detail from the fossil record involve rapid warmings. The lag is seen as a failure of the trees to establish when the climate is warm enough for trees according to other aspects of the climate record. The beginning of the present interglacial shows these effects most strikingly. The original explanation was that seed dispersal was the limiting factor. Now, soil development is identified as the fact slowing the establishment of forest. Here we should be aware of the fact that the CO_2 concentration in the atmosphere was lower than at present, although rising steeply. There is also evidence for lags in tree establishment later, around 10,000 years ago, when CO_2 was near the preindustrial level. In northern Europe there was a rapid, 1000-year-long cool interval between 11,000 and 10,000 years ago. During this interval subarctic forest was replaced, without a lag, by tundra vegetation. The absence of a lag suggests that trees were killed outright by the change, which must have exceeded a critical threshold for trees.

Waggoner: What was on the ground? Something must have been there.

Davis: In North America, spruce was replaced by temperate trees 10,000 years ago. These systems have not been studied in enough detail to learn what was competing with the spruce, and whether the trees were dying in situ, being destroyed by fires, or whether they were being replaced gradually through a process akin to biological succession. Past rapid climate changes were often accompanied by natural disturbances that speeded up the response of vegetation. In a model you can show that the lag in response of a forest community to a rapid warming or a rapid cooling might last about a century. But if you simulate disturbances, the resident trees are removed from the simulation and the change occurs much more rapidly.

Waggoner: From your knowledge, could you predict what might happen here in Minnesota? Would you get the present trees persisting well into the latter part of the next century while populations of other trees slowly expanded? Would there be a forest or would there be no forest for a while?

Davis: It depends on how rapidly the climate changes and what the threshold is to actually kill a tree. Without that knowledge, I couldn't even apply a model to simulate what would happen. If there is a slow climate change, a tree can continue to occupy territory, while its seedlings fail to persist and seedlings of more southern species get established underneath the canopy, assuming that the seeds are available. An alternative scenario is that disturbances (storms or fires) take care of the problem. There is evidence in the fossil record that disturbance rates are very closely tied to climate. So we might expect to see disturbance rates changing. In the Midwest you might expect to see a disturbance regime dominated by either wind storms or fire.

Crosson: Does anyone know even roughly what percentage of the world's forests today are managed from the standpoint of the economic gain? I'm setting aside deforestation, which is not forest management but, rather, cutting trees in order to do something else. I'm asking about managing the forest in order to maximize the economic gains. Do we know anything about that? Would you guess it to be large or small, like 10 percent or 80 percent?

Davis: In North America I think the percentage would be small. In other parts of the world, certainly in Europe, the figure would be quite high.

Crosson: The reason I ask is that if people who are interested in managing forests come to accept the implications of climate change for forestry that you have outlined, it seems to me that it would give managers an incentive to start to increase the cutting of old growth and to replace it with shorter, faster growing varieties. It would increase the economic premium on shorter-life species. If a shift of that sort occurred, this could actually improve the CO_2 problem because the faster growing new trees are absorbing CO_2.

Rawlins: Depending on what you do with the old growth.

Davis: That's right. And replacing the old trees might affect the fertility of the site. Release of CO_2 from the forest floor could be larger than absorption of CO_2 by new regrowth.
  The scenario you describe might occur as soon as property owners

perceived that the trees were not putting on wood rapidly. They would log the forests and replace them with a different forest species. The question is, what species should they replace them with? And I suppose they would be looking for species that have broad adaptability, like aspen, which grows everywhere. An early successional species such as aspen has well-dispersed seed, as well.

Clark: A lot of what Davis says applies primarily to mid-latitude forests. In tropical forests, the rates of climate change will be slower and the rate of forest growth faster. There would be at least some a priori reason for saying that the impact might be less drastic.

Davis: I agree. But the land use impact might be larger.

Clark: Right. And the point I want to get to is that one must start differentiating places in order to figure out, if one's concern is forests, whether the direct effects of land use transformation, such as arable expansion and rangeland expansion or the second order effects of climate change, are going to dominate. Those balances would be radically different, depending on which century you took and on whether you took northern or southern forests. Even for the tropical forests there is some basis for expecting that, should the climate equilibrate, you might not have an increase in the extent of tropical forests, even deducting the amount removed for land use clearing because of the wider climatic zones created. In contrast, in the northern ranges, a modest increase of land in agriculture might really put some squeeze on the boreal forests and some of the northern temperate forests.

Davis: I agree. You can't discuss it without a time frame. Vast areas of tundra might become climatically suitable for forests. Whether the substrate would be such that trees could grow is another matter. And whether the trees could establish themselves rapidly over such vast areas is another matter, too.

Sanchez: One reads about planting trees to offset the greenhouse effect. Is it correct that we may need to plant an area equal to France or something in that order to offset the CO_2 effects?

Rayner: Yes, every year.

Davis: In many parts of the world, we're cutting more rapidly than foresters are replanting. Even if we replanted to keep up with the amount we're cutting, that would be an improvement. But as Crosson pointed out, the incentive to cut and replace would become large if the trees are perceived to be growing slowly because of maladaptation to changed climate. That means many forests would be harvested. The probability that people would continue to cut faster than they replant would be quite serious unless it is countered by replanting for the purpose of restoring carbon balance.

Rosenberg: A France each year might not be enough. Roger Sedjo at RFF did some calculations (Sedjo and Solomon 1989). There are three gigatons of carbon remaining in the atmosphere or incrementing into the atmosphere each year. He calculated how much new plantation would be required to extract that three gigatons of carbon annually. Using fairly conservative annual growth accumulations, something like three-tenths of a kilogram of carbon per square meter per annum, an area about two-thirds the size of the United States would have to be planted to fast-growing trees. And if you could plant all of that at once, it would work for about 40 years. For 40 years you would take out three gigatons a year, but then your trees will stop extracting carbon. They will reach their maturity, and you will have to cut the trees and replant.

Waggoner: And store the carbon.

Rawlins: Or replace fossil fuel.

Clark: But there also are recent disturbing new data concerning the carbon that has been assumed to be going into the oceans. It doesn't seem to be in the oceans. It seems to be tied up in biosphere. So a lot of bets are off right now regarding what fiddling around with the planet's forest cover and soil will do with this carbon cycle balancing. And at least the folks I talk to right now say that we know less than we thought we knew a year ago.

Rosenberg: Okay. But no more, or not much more than three gigatons is accumulating in the atmosphere. If I understand Clark correctly, he is saying that the terrestrial biosphere probably has been soaking up even more carbon than we thought over the years because the oceans apparently have been soaking up less. There's a misappre hension that the rate of CO_2 accumulation in the atmosphere is increasing. It did for a couple of years around the last big El Niño event. But it's now down to about the same rate as it has been over the last decade or so.

Abrahamson: Between one and two million square kilometers per year of new forest will take out a gigaton a year.

Rayner: Is that for normal commercial forest or high intensity biomass?

Abrahamson: That's normal commercial forest. And, of course, as you say, that is on average until it reaches its 40 years or whatever it is before maturity. But there's something a little disturbing in that. Perhaps some of you know why the CO_2 concentrations in the atmosphere did not respond appropriately to the reductions in fossil fuel use that took place in the late 1970s and continued into the 1980s. Either the sink is being poisoned or there's a source that has not been taken into account. And that source may be increased rate of respiration with the temperature.

Rosenberg: I don't know how much you want to explore the carbon cycle, but the one other very interesting item of data bearing on this argument is that the amplitude of the annual carbon dioxide cycle at Mauna Loa has been increasing. This is reasonably well established. There are a number of explanations or contributory mechanisms, but one that cannot be ruled out is an increase in biomass in the temperate regions. This doesn't mean that the tropics are not being deforested, but the tropics exert little control over the annual Northern Hemisphere cycles of photosynthesis and respiration. In the tropics, photosynthetic activity varies very little with the seasons. But in the temperate zones, of course, we have the large annual amplitude that gets larger in the high latitudes. This increase in the amplitude of the CO_2 concentration wave must signal an increasing terrestrial biomass -- more photosynthesis, more respiration -- occurring outside the tropics. This fact, together with some of the other things we've been hearing, tends to suggest that the CO_2 fertilization effect, even though we can't prove it and we can't measure it directly, is, indeed, occurring.

Davis: I would emphasize again that in considering forest responses, the rate of change is much more important than how much change would occur at doubling or tripling CO_2. The same kind of reasoning applies to looking at agricultural responses. It's really how fast these changes occur that affect how well we can adapt to them. It seems to me in looking at forest responses, that even given the uncertainties of the GCMs and the uncertainty about what's going to happen to CO_2 in the atmosphere, a lot can be learned by looking at a rapid rate of change and seeing what that would do, and looking at a slow rate of change and seeing what that would do, and coming out with some alternative scenarios. I wonder if this is not also a useful way to approach the agricultural scene.

Waggoner: I want to go back to a point I tried to make: There's nothing wrong with cutting trees so far as carbon dioxide is concerned, and nothing right about planting trees itself. The important thing is to have rapidly growing forest rather than stagnant mature ones or bare ground. That thing seems to drop out of sight every once in a while. It isn't how many acres we plant, it's how many acres we have growing rapidly. The second issue is what we do with the wood after it's produced that matters.

Clark: Wait a minute. If you take a hectare of mature forest and cut it down and burn it up, you've just lost a hectare worth of carbon into the atmosphere. Replacing that with trees growing very fast simply to recapture the carbon you just released doesn't gain you very much.

Waggoner: If you burn those trees and replace fossil fuel, or if you take those trees and build a house, you probably have gained. That's what I want you to understand. Be sure you don't just think that cutting trees is bad or planting them is good. Think through the whole thing. What you want is a rapidly growing forest. And what you want to do is either to burn the product to replace fossil fuel or to build a house. That's the whole story.

Rosenberg: Just one more point to round out the argument. You also have to think in terms of afforesting areas that are out of forest now. Badlands in many portions of the world, particularly the tropics, may be hospitable to trees. So there appears to be many thousands of hectares on which trees might be planted. This could only be a net benefit in terms of stabilizing CO_2 levels. Afforestation, if promptly initiated, could have some effect in controlling CO_2 accumulation in the atmosphere, even within the next 20 years.

Waggoner: If they grow rapidly.

Rayner: You would want them to feed the food technology, which Rawlins was talking about, or to feed the kind of biomass program that is appropriate for the developing countries, which I was talking about.

6

Stipulations, Conventional Wisdom, and Real Issues

Paul E. Waggoner

Ruttan: Waggoner has been giving considerable thought for a number of years to the issue of the impact of global climate change on agriculture.

Waggoner: First, I will make some statements about climate change that I believe deserve to be promoted from hypotheses to stipulation. And then I'll mention some conventional wisdom about climate change that I believe merits demotion to hypotheses. And, finally, I want to talk about some issues that I think hang over the whole matter of climate change and agriculture. Stipulations, of course, are the agreements made between attorneys before a trial starts in order to get the things that everyone agrees on out of the way.

My first suggestion for a stipulation is that both a cooler climate and a warmer climate with the same water resources is unlikely. This allows us to concentrate on three future situations for any locality: (a) same climate, same water; (b) warmer climate, less water; (c) warmer climate, more water. Now, that doesn't seem like much of an advance given all the calculations that have been made over the past 10 years. But when you think about it, we have eliminated from this list of three, something that preoccupied us tremendously back in the days of the supersonic transport: namely, the possibility of a cooler climate. And I must say that I recently presented these three options, and a contrarian immediately contradicted me. But I think that we could stipulate those as the three possibilities.

The second stipulation I suggest is that a progressive 2°C warming and 10 percent drying or wetting during a half century represent reasonable scenarios. Now, a scenario isn't a forecast. It's merely a plausible view of the future that is at least internally consistent.

My third stipulation is that reliable probabilities for these three futures will be slow in coming, leaving us uncertain for a long time. In the AAAS study that Rosenberg and I participated in, we made up a table of climate change projections. In the last column is "estimated time for research that leads to consensus." For global temperature it was 5 years. I'm sure Steve Schneider would now lengthen that. For almost everything else, it was 10 to 50 years. The recent halving of the calculated warming by the British modelers was due to some changes of cloud parameters. This has thrown everything into such uncertainty that even 10 to 50 years is probably optimistic. So I've stipulated that reliable probabilities about these three scenarios are going to be a long time coming.

My fourth stipulation is that the hardest blow from climate change on human affairs will be due to changes in water resources. From the Northwestern hills to the shore of my little state of Connecticut, there is a 3°C difference in average temperature. And that doesn't really matter very much. But, they grow lettuce in the desert in California and in the suburbs of Boston, and those are quite different temperatures. It doesn't really matter. But if there's a difference in water, it makes a whale of a difference. Precipitation is the climatic hammer that's going to strike human affairs if climate changes.

Compared to the global scope of climate change, changes in water resources are fairly local. Local actions will be possible, even profitable. There are 21 water resource regions in the United States. You can act on a regional basis, but even in little Connecticut there are towns that run out of water every time it doesn't rain for two or three weeks. There are other towns that think it's just great -- they can sell water. Localities can do something about water resources, although I can't think of much they can do about climate change.

The diversity of climates where plants, animals and men survive and even prosper, indicates that we can adapt to change in climate given time. I'll go back to my lettuce example. Of course, you can grow lettuce in Boston and you can grow it in suburbs of San Diego. But it took a while to put the infrastructure in place to make it possible and profitable.

The fifth stipulation is that we can adapt to water resource differences, but it takes time. Parts of the New York City water system in use today are over a century old. I'm not sure how long TVA was in conceptualization or preconstruction, but probably it was something like 20 years, then it took another 20 years to complete the system.

My sixth stipulation is that extremes of frost and drought have more impact on affairs than do averages. The amplitude and timing of annual cycles of temperature, moisture and runoff have more severe consequences than differences of annual total precipitation or average temperature. Amenities like recreations, scenery and wildlife, and especially anxiety about health, compete with surprising strength against bread and butter issues. I'm accustomed to limitations on fertilizers and pesticides. Nevertheless, I was surprised to hear from a Russian last winter that environmental concerns in Russia were quite capable of stopping development long before they ever had Peristroika. It's really remarkable that in a country like that an environmental issue can stop a large development project. And wonder of wonders, I think the most extraordinary thing I read this week is that the environmentalists have presumed to stop the Israeli Air Force from using a piece of the desert for bombing practice. Now, if that doesn't show the strength of environmental issues, I don't know what could. It's good, but it also justifies a statement I heard from a person who said he feared more irresponsible acts to prevent climate change than he feared climate change.

My seventh and last stipulation is that an act, in the end, always costs. Think of policy actions as investments. I have a favorite question I like to ask: How would you invest your own money to make 10 percent or more per year on your insider's knowledge about climate change?

Well, now, those are some things I would like to elevate to stipulations. What about conventional wisdom that might be reduced to hypotheses? These, like my stipulations, are questions for research. One piece of conventional wisdom is that "waiting will only drive up costs." That is true only if you don't have to pay any interest and if we have dumb farmers. Otherwise it may not be true.

A second bit of conventional wisdom is "there are only losers from climate change." I don't need to talk about it. We've agreed that that piece of conventional wisdom won't stand up.

A third conventional generalization is that "anticipated changes are unprecedented." In fact, during a recent 30 year period, the range of annual precipitation in my temperate state of Connecticut was 28 percent below to 38 percent above the mean. And during the most recent 30 years, annual precipitation fell 29 to 39 percent below the mean in Bozeman, Montana; Columbia, Missouri; Pensacola, Florida; Rockville, Indiana; and Forks, Washington, which has over a hundred inches a year. It fell 74 to 87 percent below in Childs, Arizona, and Indio,

California. So changes far greater than the commonly specified 10 percent are regularly encountered. The evidence is not at all clear that the effect of greenhouse warming will be the increase of climatic variability.

A fourth bit of conventional wisdom is that "cutting forests always increases the CO_2 in the air." But mature forests fix no net CO_2. Therefore, cutting for lumber or firewood and replacing with a rapidly growing stand or crop will reduce CO_2 concentration. But when it is harvested for lumber or firewood, it will again release CO_2 into the atmosphere.

A fifth bit of conventional wisdom that I have often encountered is that "genetic engineering will save us." It is premature to begin to design crops for anticipated environments. The logical procedure is to continually adapt crops to the climate as it evolves. Depriving conventional agronomy of research support to feed anticipatory research, and betting all our chips on an uncertain future, doesn't seem to be smart to me.

A final conventional generalization is that "the poor will suffer most." The proposals for stopping the possible warming also may prove costly to the poor. Some very explicit and careful calculation or analysis needs to be made before we accept that statement.

Now, let me turn to some issues that hang over the whole matter of climate change.

The first one is, "Why have we failed to implement so many well-known and seemingly sensible suggestions?" These include energy efficiency, water efficiency, and a very long list of others. Let me read a statement from Helen Ingram, "Just as surely as solutions are sought for problems, solutions go shopping for problems. When an emerging problem lends additional credibility to an already developed policy proposal, the proposal is likely to be attached to the problem." The climate change issue, of course, is attracting all sorts of well-known solutions. It might be helpful to ask why they haven't been implemented. If there are good reasons, then let's not fill up our reports with them.

One thing that will increase the possibility of a sustainable agriculture is investment in monitoring and research. We'll surely say to do that. One criterion for investment is the net utility of an investment relative to the effectiveness of the remedy. But devoting more money to research without any impact on the problem will decrease net utility

Increasing the utility of research is important for agriculture, and it is absolutely crucial for the good name of research.

There are some very severe obstacles to interdisciplinary research. Research in separate disciplines is not wasted, but it doesn't get directly at solutions. Patrons who want to get results from research on climate change would do well to reward rather than discourage interdisciplinary research.

How does a thoughtful individual factor in climate change? Groups like ours always recommend that the water system managers of American or the seed corn producers of American need to consider the implications of climate change. Well, imagine them trying to do that. What would they do? One frequent response is to build in margins of safety. It doesn't take many brains to do that but it costs money. A real contribution would be to say exactly how to factor in climate change. "If numerous unmanageable alternatives get dumped into the deliberations, participants may decide the subject is too complex, the problems too numerous, and the alternatives too overwhelming and turn to more manageable issues." So I think there is a good tactical reason for us to learn to sort these proposals and get rid of some of them.

A way around the hard job of sorting these things, of course, is to find three or four that are so important we don't have to think about anything else. The important effects of climate change are those that have a highly elastic response, in the sense that elasticity is used in economics. John Shaake has found, for example, that the elasticity of water supply for a change of precipitation on the east coast of the United States is about 2. It rises to 4 and even more in western Texas.

Ruttan: What does that mean?

Waggoner: It means that if you get a 10 percent change in precipitation in Georgia, you get a 20 percent change in runoff; if you get a 10 percent change of precipitation in Texas, you get a 40 percent change in runoff. So if you have a very high elasticity like this then you're on to something very important in climate change. It's worth concentrating on. The other thing is that the system is very nonlinear. Moist weather makes corn grow until a fungus intervenes, and then the plant dies. Cool weather may be good for a crop until it goes below 32°, then the crop is killed.

This is what makes the changes in the extremes so important -- say the probability of drought below a certain amount of precipitation (Waggoner, 1989). That has a very high elasticity. If the mean changes by, for example, 10 percent, the probability of drought in the extreme may change by 40 percent.

Gary Yohe (1990), whom some of you know, has shown how to make good use of these nonlinearities or thresholds. Consider the issue of flooding over a levee: The sea level isn't important in itself, therefore, you concentrate on the time when the sea goes over the top of the dam or the levee. So instead of a frequency distribution of sea-level depth in, let's say, 2020, you concentrate on a frequency distribution of the time when the sea goes over the wall. Thinking in that way, you incorporate things about rate and about nonlinearity, and I think it is an advance in knowledge for which Yohe is to be praised.

Well, this ends my statement of the stipulations and the conventional wisdom as well as my list of these great issues that I think hang over everything.

Davis: Concerning conventional wisdom: you were challenging the idea that the changes will be unprecedented. And you're challenging them by giving the range of variance. Shouldn't you assume that if the mean annual climate gets warmer, the variance envelope around the mean would stay the same?

Waggoner: All I meant was that we often encounter changes bigger than the ones that we envision for climate change. Then, when we make statements to the effect that we've never seen anything like this before, it discredits the effort.

Davis: I don't think it's irrelevant, though, because a severe drought that persists for two years has a much greater effect than if it only persists for one. The frequency of drought years becomes critical for forest systems and, I should think, for economic systems as well. I don't think that we have in our lifetimes experienced climates such as we're visualizing. We did in the geological past. But the natural vegetation in the past contained very different distributions of species. During the last interglacial, for instance, sea levels and CO_2 concentrations suggest that the climate was warmer then than today. This suggests that we may see major changes in the natural vegetation given a mean temperature that is higher than what we now experience. I don't think that range of annual variability actually suggests that we've seen these things before. That range of annual variability fits better with your statement that it is the extremes that are important. I certainly agree with that.

Abrahamson: Of your stipulations the one I don't much like is in 50 years. You have to be awfully optimistic. We'll have to be lucky on the scientific end, on the uncertainty of the science, and we will have to be pretty vigorous in terms of our policy response to decrease emissions. Both of those things would have to happen. It's possible but highly unlikely.

Waggoner: I said it was a reasonable scenario. But, in fact, my stipulation was that we're not going to know what it is for a long time. Within the possibilities I think it is a reasonable scenario to think about.

Abrahamson: I would at least double the assumption about the equilibrium warming to which we are committed.

Clark: That's a little high then for the presently published consensual median estimates. But it's close enough.

Chen: Can I make one response to Davis? Jesse Ausubel's point on whether or not this is precedent or unprecedented, was based partly on the CLIMPAX (Climate Impacts, Perception, and Adjustment Experiment) work (Karl and Riebsame 1984), a research project sponsored by NSF, which did at least look back at the historical record and look for large regions to see whether there would have been significant climate changes that had persisted for some length of time. There were examples from the midwest, I forget which states, where there had been as much as a two-degree change in the mean from one decade to the next. And people didn't seem to notice the difference.

Rosenberg: I'm sure they noticed the difference.

Chen: No, they really didn't. There was nothing in the popular press. There were no expressions of concern about it being warmer this decade than the last.

Rosenberg: The very big difference was between the decades of the 1930s and 1940s. And people surely noticed that difference. But CLIMPAX did identify limited areas in which other, less dramatic decadal anomalies in temperature and rainfall occurred. It may be that these weren't large enough changes to cause major impacts on society. That may be why it didn't notice.

Clark: I think a way to sort out a lot of the confusion is to note that it is not the rate of change, per se, that's unprecedented, it is the combination of a high rate of change sustained over a long period. If you plot the records of the last 160,000 years, or of the last several million years -- the combination of rates and durations -- you get a red spectrum, which is the standard distribution of climate-type noise, showing that the biggest total fluctuations come from events that are of very slow, very long duration. You get a lot of very large changes that persist only a very short time, like noon to midnight, up at the other end of the spectrum. What is interesting about the climate change from greenhouse scenarios is that the sort of changes forecast up to the present have a fair amount of precedent in the historical record. But if modest rates of change are continued for another 30 or 40 years into the future, much less accelerated, you just get into an area of the rate/duration space where there are no historical observations. So once again, the statement that it's unprecedented, without specifying spatial scale and combination of rates and durations, is a non-argument simply because you're not specifying enough of the dimension to have something that data could refute.

Waggoner: That's exactly right. You have made my point better than I did. I think another useful concept is Yohe's idea of looking at the time when the threshold has passed. One more piece of conventional wisdom, which should be reduced to a hypothesis, is that we will know the impact of a future climate on agriculture by making calculations using present crop varieties. That's the dumb plant breeder's assumption. It should be eliminated from the conventional wisdom.

Abrahamson: I have one question for you and also for the group. When you said variability will increase, do you mean interannual variability in weather events?

Waggoner: We haven't calculated that. We don't know that.

Abrahamson: That's my impression. There's no evidence either way.

Waggoner: Yet people will say variability will go up, but there's no basis for that statement.

7

Agricultural Impacts on Climate Change

William C. Clark

Ruttan: I want to ask Clark about one thing that seems to be falling through the cracks. We have talked quite a bit about the implications of environmental change, both global and local, on agriculture. But we haven't put on the table yet very much about agriculture's contribution to either the global or the more location-specific environmental problems. Clark has been heading up a committee that is specifically charged with looking at some of those effects.

Clark: Ruttan is referring to a committee of which Davis is also a member. The National Academy of Science is trying to outline a research plan for the U.S. Global Change Program. A group composed of natural scientists, plus a few of us who once were natural scientists, went through an exercise in which we tried to focus on the intersections among the classic disciplinary areas of research. The climatologists can define what climatologists want to do. Even the ecologists could almost define what ecologists want to do. But the difficulties in research planning always have been at the interfaces. What do the climatologists need from the ecologists to get on with their work? Research that the ecologists might well not do as part of their internally driven agenda turns out to be essential for getting on with the climatology or the atmospheric chemistry.

As part of that exercise, one of the questions asked was what do all the science disciplines -- climatology, earth system chemistry, ecosystem dynamics, and so on -- need to know in terms of the human forcing functions that are pushing perturbations in the global geosphere/biosphere system? As you might expect, they identified a whole set of issues that had to do with industrial and energy emissions, which are not primarily our concern here. But their second big class of categories was things that result from land use change in general and agricultural activity in particular. I guess what you're asking is that I just run down, as best as I can recall, what our answers were.

There will be a couple of categories. The most obvious is which land-use change activities are resulting in emissions of chemicals, primarily gases, that contribute to changes in climate and/or changes in tropospheric chemistry. (Those being two of the dominant global linkages now on the research agenda on global environmental change, which as noted earlier, was dominated by atmospheric chemistry and climatology.)

The first thing one does is to identify the set of greenhouse gases and ask which are mediated by land-use transformation activities. Carbon dioxide, one of the major greenhouse gases, is certainly affected by land-use changes, primarily through direct forest clearing; that is, clearing high biomass standing stock, the combustion of that material, its release to the atmosphere, the plowing of soils, and the oxidation of those soils resulting in the release of carbon dioxide.

A second is methane coming out of agricultural activities via two routes. One is anaerobic production within ruminants. When I was an undergraduate, I thought it was an interesting choice of words -- the most interesting human contribution to the planet's atmospheric chemistry. And indeed there is an interesting amount. It turns out not to be a big number relative to other numbers in the accounts, but it's certainly been rising of late. There are more ruminants around than there were 150 years ago.

Almost certainly a large agriculturally related source is any land area that is wet enough for the anaerobic route gives you CH_4 instead of CO_2. This clearly happens in rice cultivation and in other wetland-like operations. It also happens in seasonally flooded areas and in very damp soils. One of the great difficulties is that the carbon evolution can switch between an aerobic and an anaerobic pathway very rapidly. It means that the emissions patterns are extremely spotty. You can be getting methane out of a system one day and CO_2 the next day. So it's a very difficult thing to sample or to understand the sources. But it's equally clear that increases in irrigated areas lead to increases in methane evolution, unless those systems are simply replacing natural wetlands.

As far as anyone can tell, there is no ozone source out of agriculture except from internal combustion engines. Nitrous oxides, which are a significant greenhouse gas in that they have extremely long life times, come out with great high uncertainties at about 50 percent from fossil fuel combustion and other industrial combustion processes. The other 50 percent comes from biomass burning, soil fertilization and cultivation of natural soils resulting in some rather bizarre chemical pathways that involve N_2O. The difficulty is that these numbers are not well known. N_2O has been a very difficult gas to sample. It is well distributed because it has a very low atmospheric lifetime, but it is difficult to detect at the extremely low levels that exist in the troposphere. Only very recently have sufficiently robust sampling technologies been put in place to begin to get a clear picture. But nitrous oxide is certainly going up. The numbers you will see around in most literature are now somewhat suspect because the portion attributed to fossil fuel burning is in doubt. But over the long run those nitrous oxide sources are something that people will wonder a lot about in terms of where they're coming from.

That said, the second class of major chemical issues are those that result primarily from biomass burning, such as slash and burn or burning the waste materials in a cropland or forestland after clearing or harvest. These emit a complex set of gases. Some I've already mentioned, but some are much more complicated: low molecular weight hydrocarbons, aerosols, small particles, soot particles, and the like. There is a fair amount of sulfur in it. You may have seen the recent public reports that some fairly significant acid deposition damages were being measured in what used to be called the Ivory Coast of Africa, far away from any plausible sets of industrial sources. They were apparently traced back to the quite extensive burning of vegetation. A combination of the moisture conditions and the sulfur content of vegetation were quite capable of giving you sulfur aerosol rain downwind from the burn off. The point is that sulfur deposition has been appearing in places that nobody was expecting.

Other effects of this very complex chemistry of incomplete biomass combustion have been a whole set of photochemical smog-like phenomena. This is occurring even in remote areas. You get some very bizarre air chemistry that can stress all sorts of things. There may be impacts on vegetation and conceivably, eventually, impacts on human health. The next major category is the water budgets of the earth; that is, the land-atmosphere fluxing of water, turns out to be very strongly mediated by the vegetation cover. This has been one of the areas left out of the first generation of global climate models. The radical differences in the ability of the vegetation surface to pipe water from the ground into the atmosphere among bare land, a smooth field, and a forest or brushland has been getting a lot more attention and is now being made a parameter in the next generation of global climate models. So there's a lot of research going on in that area that builds on a long tradition of agrometeorology studies, but they have not, until recently received much attention to scaling them up to, say, scales of tens and hundreds of kilometers. Finally, I guess I merged into the last issue: the surface properties themselves. Obviously, land-use transformation changes surface properties as they affect the fluxes of water. They also affect the incoming solar radiation budget -- different degrees of reflectivity and different degrees of wind scouring.

The natural science communities have been asking, what can you tell us about plausible internally consistent patterns or scenarios of land use transformation as they affect these various transfer agents, chemicals, physical balances, and so on, over the next 50 to 100 years: They're not looking for predictions. They're looking for sets of plausible reference scenarios. What would agriculture look like, in terms of its methane emissions, N_2O emissions, surface cover changes and the like? What would radical alternative patterns of agriculture, these appropriate technology or sustainable versions or whatever, look like in terms of those transfer parameters?

That is the agenda that is being pushed. The missing agendas tend to be the ones that are not directly atmosphere and climate related. They have direct implications for the diversity issues I spoke of earlier and for the fluxes of materials and chemicals into the water system. They are acknowledged, in passing, in terms of the phosphorous budget and its involvement especially with carbon sequestering in the deep ocean. How much carbon and how much phosphorous is being flushed down the major world river systems? But they are very much second tier concerns at the moment in the global change program.

As far as our committee's work, we have simply bowed to a lack of demand pull and relegated those waterborne and direct biotic effects to relatively back-burner status simply to get the atmospheric chemistry and surface properties questions answered first.

Allen: I have two questions: I didn't hear mention of the role of termites in methane production and whether there is, in fact, a large unaccounted portion of methane generation.

Clark: There is no question that most but not quite all termites and a whole bunch of other creatures do produce methane in their guts. They're doing anaerobic fermentation. They don't have much choice in the matter. There was, several years back, one of these elegant little exercises where, having measured the evolution of methane from one cubic centimeter of termite land, you then try to figure out the scaling factor of how many such cubic centimeters were there in the universe. There's an error term in that estimation. Depending on which ends of the possible range you pick, you can turn the world into a methane planet or methane becomes an insignificant source. The present view is that the methane budget is unbalanced. The atmosphere isn't getting rid of as much as it should be. Something is happening at land surface that isn't the termites. The termites are in there as a source term of unknown size. Most recently, some very elegant isotopic analyses have suggested that a larger fraction of the methane is of fossil fuel origin; that is, coal mine surfaces, incomplete combustion -- a whole bunch of things. Very old carbon is now being combusted incompletely or there is more methane leakage from old carbon than had been thought to be the case. The atmosphere people said that couldn't possibly be true. But I think right now almost anyone who knows anything about the methane issue can give you a good argument about why it goes one way or another.

In the December issue of *Biogeochemical Cycles*, Ralph Cicerone (1988) gave an absolutely gorgeous review of the topic from the point of view of laying out the constraints on the global methane budget: what we know about the isotopic measurements, the known sources, and the known sinks. Instead of taking a central estimate and putting an error term on it, the writer came in from the outside and asked what is the space within which the right number has to fall? It's a very clear and systematically written piece, now somewhat superseded by some of the new isotope analyses, but the structure holds up quite nicely.

Rawlins: Has dust from wind erosion been considered a contribution from agriculture?

Clark: It comes up any time you ask the question of mediators of mesoscale climate over periods of months or years. We clearly get transcontinental movements of significant quantities of dust. I don't think the Reed Bryson notion that there was sufficient mobilization of such dusts or dust-like aerosols to significantly increase the reflectivity of the atmosphere, that is, to increase the albedo and not let as much sunlight in as expected, thus pushing us in a cooling direction, has borne out. It's not that it couldn't do it, it's just that the masses involved simply aren't there.

But the whole aerosols and dust issue, as many would say, has been given very short shrift in the global and continental scale atmospheric chemistry arguments. If we were around this table in Germany or the Soviet Union right now instead of in the United States, there would be a lot more discussion about this and a lot more argument that a significant fraction of some of the tropospheric chemistry, and even the mesoscale climate effects we're seeing, are due to such things.

Rawlins: Where do you see the research priorities in agriculture? Where do you see the important gaps that need to be filled?

Clark: One thing that the committee has come out with is to say that we face a very odd situation in our ability to talk about internally consistent long-term scenarios of human activity on the planet. The demographers and energy people are perfectly willing to give you hundred year scenarios. Despite the excesses on either side, I think most people in the environment and development debate would argue that that's a good thing. Both fields have matured sufficiently that there is good critical peer review. One can make a distinction between sloppy work and solid work without falling into the trap of believing the numbers. The odd thing is that when you come to agricultural change in particular and land-use change in general, the long-term studies go out to 2000 or 2010. Most are static, for example, an FAO carrying capacity study. It has been virtually impossible to stimulate work on the principal driving forces in large-scale persistent land transformation that would show up over scales of hundreds of kilometers and tens of decades. I'm not concerned about the high-frequency back and forth this year and that between grazing land and cropland, but the larger more persistent changes. What are the varying roles of demographics, or prices, or demand, and so forth? What, if anything, can we say about the constraints and the determinates under which these patterns emerge, especially as they relate to some of the land-use transformations that have shown to be most important for the chemistry and climatology issues I've been discussing?

We need to challenge the agricultural economics community because nobody else seems even remotely placed to do the work. You must have people who are dealing with long-term processes, the kind of thing Ruttan talks about, long-term technical substitutions, and long-term demographic transitions and what they mean for these land-use issues. Try to challenge them by doing the first cut, first draft global model of land-use changes for five or ten decades into the future. The nice thing about that, of course, is that the model isn't the objective. Identify a dozen or so key processes that are the determinates of transformation and explore them using historical and cross-sectional data to get a debate going on the decade-to-decade changes. That's where real research gets done. It's an area in which all sorts of both cross-sectional and historical studies have shown good work can happen but hasn't been done. Try to integrate that work, then, with changes in major cropping zones. That is the kind of task that I think would bring the agricultural development community and the global change natural science community together. It would give them one place of common contact which I think would be very good.

Rayner: As far as identifying the important processes in long-term land- use change, my own institution at Oak Ridge National Lab is just starting a three-year program of research in that area. We would welcome any suggestions or inputs.

Rosenberg: If you take the relative size of RFF and the relative size of Oak Ridge, the effort that we are beginning is about on the same scale. We've agreed to do a one-year survey with support from the Japanese Institute for Energy Economics on emissions of non-CO_2 biogenic greenhouse gases (CH_4 and N_2O). To do this we will judge the validity of current estimates of land-use change. The validity of methane and nitrous oxide flux measurements will also be assessed. We have not yet actually started, but we're gearing up now to do it.

Allen: You didn't mention ammonium relative to animal units and the use of anhydrous ammonia as a fertilizer. Are these significant to global climate?

Clark: Anhydrous ammonia isn't a climatically active gas at all. It has a very short atmospheric resident time: a matter of days. Its transport distances as ammonium are not sufficient to get it involved in global change research. Now, at mesoscale, obviously there are places where it gets quite involved with nitrogen oxide or nitrogen acid deposition patterns through various chemical pathways.

Rayner: It's also important with respect to fertilizer manufacture. One-third of all the energy use in U.S. agriculture is in ammonia manufacture. In addition to the energy use in the manufacture you also have natural gas. The carbon from that usually is in some kind of urea form. When that is subsequently released it contributes to atmospheric carbon.

Clark: Let me switch hats and stop talking about what the Academy's global change committee is doing. What I would hope this group would support is research on how we deliver nitrogen to the crops we want to grow. What's the difference between the amount of nitrogen that the farmer applies and the amount that ends up in a crop? That is a classic systems mass balance study that needs to be done: not at a global scale, but certainly at a couple of hundred to thousand kilometer-on-edge scale. We should begin by looking at the agricultural nitrogen budgets from a point of view of what the chemical products are, what's going into the surface water system, what's going up into the atmosphere, what's going down deep, what's in the plants.

Rayner: Is everybody actually familiar with the different potencies of the various greenhouse gases that Clark has been talking about? I assume you were talking there about the weight of gas emitted, right? You were not talking about the relative contribution to force and effect?

Clark: True. They have very different numbers, and CO_2 is the weakest. For what it's worth, an analysis by World Resources Institute of the next 30 to 40 years of forcing global warming attributes 13 percent of that forcing to agricultural activities. Within that, 3 percent is attributed to carbon dioxide, 8 percent to methane, and 2 percent to nitrous oxide. Those would not be the numbers if we talked about volume or weight emissions.

Rawlins: One reason I asked what agricultural research priorities should be is that the figures for gaseous emissions from agriculture frequently seem to be the residuals left over after the emissions from energy production and industry have been estimated. Do you think agriculture should take the responsibility for assessing these emissions? One of the weakest numbers seems to be the relation between nitrogen fertilizer use and nitrous oxide emission. Do we know that this is the source?

Clark: No, we know that N_2O comes, among other places, from nitrogen fertilizer. You can do it in the laboratory and pick it up in the field. It happens. It's a matter of how much of it happens, where and when. I think there is a genuine opportunity for the agriculturally based research community to get involved in improving these numbers. What would probably be a lost opportunity would be if the research community retreated and said, okay we'll do this within our own community, instead of moving in and playing an active role in the existing cross-disciplinary effort that has been unable, with a very few exceptions, to get good sustained cooperation from the agricultural community. What it means is that they do the best job they can with the collaboration they've been able to get. And I don't know a single group doing these budgets that would not be delighted to have very substantial interactions with the agricultural research community.

8

Food Systems Approach

Robert Chen

Chen: I was at a meeting a couple of weeks ago where Bob Dickinson, a climatologist for the National Center for Atmospheric Research, made the following comment: "Climatologists know a lot about the climate, but they also don't know a lot." And that's reflected in the comments about how long it will take to have the answers -- climatologists just need to do more research for five to 10 years and we'll have all the answers. Dickinson argued, "No, we're not. We're going to have as many new uncertainties coming up as we will solve in the next ten years." My view is that, in the impacts area, we actually *know little* and, in fact, we *know little about how little we know.* The extreme positions on whether impacts are severe or not are based on very little hard knowledge.

If you look at a simple system in which there is an activity that you're worried about, such as climatic change, it helps to think in terms of the principles we learned in basic calculus. There is a term, the total derivative, which is, say, the change in agricultural production that results from the activity you are concerned with (e.g., fossil fuel use). This limited variable system has a total derivative reflecting how the change in the activity affects the climate.

There is also a partial derivative that reflects what climate change, holding all else equal, does to agricultural production. Another term of interest -- though maybe not to climatologists -- captures the benefit of fossil fuel use to agricultural production. These make up the total derivative, so from a societal viewpoint it is important to remember that we must also worry about the benefits from fossil fuel use in agriculture.

Certainly one important thing is that significantly large negative effects on agriculture due to climate change must be established in order for the climate-agriculture connection to be important from a societal viewpoint. If this derivative is zero, then, since the effects are multiplicative, there is no net effect on agricultural production. So there is a need to establish that this is a large negative effect before one can proceed in saying that the adverse effects of climate change somehow outweigh the benefits of using fossil fuels in agriculture.

In my work at the World Hunger Program, we've come up with two approaches to looking at some of the impacts. One, you might call a "food systems" approach. Most approaches to climate-impact assessment in agriculture have not taken a broad, food systems approach. They've actually only focused on single issues on the production side, such as yield impacts, effects on inputs, some attention to pests, and not much else. Let me discuss a diagram that I've been working on for the purpose of rethinking the issue of food waste in the food system.

Most of the things that Clark included in his list are in my input list. Clearly, you have to worry about the effects on inputs and not just land use and yields. In Africa, for example, one of the major constraints on land use in agriculture is the occurrence of waterborne diseases that prevent human occupation. That's something that will be responsive to climate change. It may, therefore, be a different pathway for affecting eventual food consumption. It doesn't relate to the demand side at all. Energy use is also a big factor in agriculture. Fertilizer is basically an energy use issue. Rawlins mentioned the possibility of using cellulose, for example, to feed livestock. That would introduce a new component into the food system. But another issue is the waste that goes into feeding livestock. If one could increase the efficiency of feed conversion, it would probably far outweigh the effects of a 10 or 20 percent difference in corn and soybean yield.

There are lots of points of vulnerability which I won't go into. Certainly one interesting one is the whole issue of waterborne disease. There is also the issue of disruption of the food system itself. In Africa, for example, 2 percent of the population are refugees of some sort or another, and there is some potential for that to increase drastically. This has led to regional and sometimes national-level disruption of the social infrastructure that keeps people fed and housed.

A related issue that was mentioned earlier is temperature and precipitation effects. Rosenberg mentioned that people have looked at CO_2 and salinity, CO_2 and drought stress, and CO_2 and other factors. But what people have not looked at in any great detail is the whole set of cumulative system effects. There have been few synergistic studies of what happens to crops or forests when there's ozone, when there's air pollution, when there's a whole range of climatic stress.

Waggoner indicated that a 10 percent increase or decrease in solar radiation over the growing season will result in a 10 percent increase or decrease in yield. But you don't see those kinds of estimates worked into agricultural impact studies, principally because the modelers are not prepared to release results on insolation effects because the cloud parameterizations are so weak.

A second issue that has not been addressed to any great degree is the possible connections between climate change and some of the other environmental changes. They may, in fact, have synergistic effects. In addition, there are parts of the system that modelers do not normally think of. I put agricultural research in that category. Agricultural research will have large consequences for the impact of climate change on agricultural production. But climatologists, and others who have traditionally defined this as a climate problem, have generally ignored agricultural research.

Finally, an alternative to a foods systems approach is to think more in terms of existing societal relationships and how things like food shortage, distribution of food and other factors interrelate. This diagram is too complicated to explain in detail, but it represents our current thinking of how to deal with hunger. Risk relations vary at different levels of spatial aggregation. Social relationships become important because the issue goes beyond the capacity to produce access to food. Greater food shortage increases the numbers of people who have inadequate access. But it certainly is not a one-to-one relation. Inadequate access can occur because of other factors in society. Even within the household, individuals have different access to food and different requirements.

This kind of framework allows you to think not of flows of calories, but rather of a hierarchy of risk. Changes in climate which may affect food shortage may also have other implications as its effects are transmitted through the system. This gets to the issue of whether there are absolute winners or losers. Perhaps there will be no absolute winners or losers, but certainly within a system in any particular region, there will be relative winners or losers. This may be more important to popular views of the problem than the absolute level of risk.

The effects of climatic change may well occur through other pathways, perhaps through changes in economic relations within society. Policy actions may modify economic relation within society and change the degree to which, for example, hunger persists in a world subject to climate change and to policies to prevent, adapt, or mitigate climate change. This is still preliminary, but it does provide an alternative way of thinking about these impacts that is different from the old "let's do a scenario and add up the costs and benefits."

Waggoner: I would like to ask if Rayner and Chen can tell us how we can use some of these techniques to sort through the proposals for either stopping the climate change or adapting to it to see which is or is not reasonable.

Chen: Well, one issue is that the impact studies are so limited that even if you think there may be some CO_2 benefits, there is still a risk of catastrophe. That suggests a more conservative strategy in terms of prevention.

Clark: A simpler and stronger answer is that you should never fund a single additional impact study which has the "dumb farmer" in it. No study that is looking forward 50 years and does not incorporate a mechanism for behavioral and technological response, should ever be funded or considered again! It simply creates a set of numbers and pictures of a world which is inconceivable.

Ruttan: You had better go a bit farther and rule out not only the "dumb farmer" but, also, the "dumb plant breeder" and the "dumb animal nutritionist." We should assume that the research system has some capacity to respond to the changes in the environment. And if any body doesn't believe that, I'll refer them to Hayami and Ruttan (1985) on "induced innovation".

Clark: The point is we have enough of a tradition in assessment and response studies for multiple-decades time horizons. It is not even a useful first approximation to do the study as though those responses didn't happen. Yet 90-plus percent of the studies out there -- even the best of them -- are systematically biased in such a way as to make the results or the consequences look much worse than they're going to be.

Rayner: I think it's important not to assume that the smart farmer or the smart plant breeder is necessarily going to make things better. They could make them a lot worse.

Rosenberg: I agree with your premise that the "dumb farmer" scenario is nonsense. The no-adaptation assumption is silly. However, the very fact that so many studies have used this assumption virtually requires those of us who want to go beyond that to use it as a starting point. Policy is, after all, being proposed in Congress and in international fora based on the results of studies that have used "dumb farmer" scenarios.

Crosson: In response to Waggoner's question, one area badly needing economic research seems to be this question of the gains from conservation. Are the technologies to increase efficiency in energy use economically viable now? If so, then the question economists always ask is why are they not being used? My point is that if these gains in energy efficiency are as large as asserted, and are even potentially economically viable, then we need to know it. We need to answer the questions, why aren't they in use? They would undoubtedly require some institutional changes, but the greater those institutional changes, the greater the cost. Institutional changes don't come cost free. But if they are, in fact, available at low cost, then all the problems of the uncertainty about climate change becomes a non-problem because it's only sufficient to show that there is some cost to not doing something about climate change. If we can deal with that at very low cost, then it doesn't make any difference whether we don't have good estimates of the future costs. I've been arguing that the RFF people in energy ought to be paying more attention to the economics of the conservation strategy and the energy efficiency strategies.

Rayner: You can push that even further back to the question of what is the starting point for impact analysis. One concern I have is the extraordinary degree of technological precision of the analysis in terms of material flows and emission rates combined with very superficial attention to the institutional structure and its implications.

The whole question of the introduction of a dynamic decision maker to impact analysis makes the comparative static analysis misleading. We've tended to rely too heavily on late 1960s and early 1970s systems-- theoretical and flow -- modeling approaches and to ignore active choice making. Until we can start to actually integrate the active choice making into our impact analyses, we're always going to have the problems that Clark just said he would like to see excluded from modeling.

Sonka: Waggoner asked about what tools and techniques exist to help in prioritization. The tools and techniques used in the studies I will be discussing were useful and cheap. I'm struck with the odd choice of

80

words. We distinguish between impact analysis and science. I don't want to sound defensive but we don't put that kind of money into studying the science relevant to impact analyses that it deserves. When an issue comes along you pull the old models off the shelf.

Waggoner: Perhaps the reason is that you haven't convinced somebody that you can do better.

Rosenberg: I have the answer. First of all, there have been any number of studies about the impacts of a climate change on agriculture. The EPA study illustrates the inadequacies of the "dumb farmer" assumption. The studies show that yields go down. Yields go down because heat units go up. In the plant-growth models used, growth rate is determined by the accumulation of heat units. Hence the plant runs out of time. The crop stops dead because it reaches its heat unit limit weeks early so that the time for accumulating photosynthate is curtailed. In real life, farmers perceiving a warming trend and observing that crops are maturing too early would begin to plant earlier or otherwise change their management practices.

Ruttan: The plant breeders would decrease the days of maturity from 110 to 100 days.

Jones: They already have the varieties. They would just pick different months to plant.

9

Modeling the Social and Economics Effects

Steven Sonka

Ruttan: Sonka has been attempting to evaluate the social and economic implications of the projections from the climate models. His perspective on what he thinks we can learn from them and what we need to do should be helpful.

Sonka: What I've been looking at is the studies that have been trying to measure the social and economic effects of climate change on agriculture. About six months or so ago, the Organization for Economic Cooperation and Development (OECD) commissioned me to undertake this review. I was asked to look at the methodologies that were being used to assess the potential social and economic effects of climate change on agriculture.

We tried to do three things in the study. One was to just review the empirical studies to find out what has been done. We were able to draw on Martin Parry's work at IIASA quite a bit. The second objective was to critically assess the methodologies used in the studies, and the third was to suggest improvements.

In another incarnation, I teach in the area of management. And one of the things that we talk a lot about in management, particularly in the strategy area, is that you should spend some time thinking about whether you are doing the right thing as opposed to spending a lot of time thinking about whether you are doing things right. The bottom line from my review of the studies is that we've been spending a lot of time worrying about the correct way of doing the studies, but we haven't worried very much about whether we are doing the right studies. We are not providing the information that societal decision makers can use to make these very, very difficult choices.

I'd like to take just a few minutes to review what we did, then talk about where we think some of the key deficiencies are, and then relate them to our earlier discussions this morning, particularly Chen's presentation.

One of the first things I tried to do was to move away from climate change as the primary issue. From the viewpoint of society, or of societal concerns, climate change is not the primary issue. I stress this because in the studies that have been done, almost all have taken climate change as the central issue. As we thought about the basis for society's concern about climate change, it seemed to us that food security was a primary issue. Food security has local, national, and international dimensions. Thus the dynamics of food security are the central concern.

Furthermore, society is concerned with the production and consumption of food and the economics of that system. Whether an area is food-producing or not, food consumption is a major economic activity. Climate change has the potential to affect food consumption in ways that may be as important from an economic perspective as its impact on food production.

A third societal issue is that of investment in agricultural infrastructures. Our concern with the implications for agricultural research is one aspect of the issue of agricultural infrastructures. We need to think about the implications for public and private investment in the infrastructure.

In our review, we looked at 19 empirical studies. We looked at 17 different characteristics. I won't try to talk about all here. But they fit fairly neatly into three categories. The first relates to the source of the climate change and how the climate change process was being modeled. Harking back to some of Rosenberg's comments, almost all were generated by Global Circulation Models. That was the underlying source of the climate change projections.

Two things in this area were particularly troubling to me. One was the instantaneous change approach used in some studies: that all of a sudden we wake up and the climate is different, nothing else is different, but the climate is different. From a societal or from a policy maker's viewpoint, I think it is very hard to understand such a situation. I don't mean understand it in the intellectual sense, but of what to do in a bureaucratic and political sense. That's a very nebulous kind of information.

A second concern is the very limited analysis of the uncertainty associated with the climate change process. Every report goes to great length talking about the uncertainties of climate change. But when it comes time to do the modeling, they essentially ignore uncertainty. The analysis rarely goes beyond a little sensitivity analysis -- typically rather naive and not very meaningful. I don't want to be too critical here. I view the methodologies as evolving. If I were doing these studies I would probably have done the same thing over the last three to five years. Our criticism is directed toward how to conduct the next set of studies.

We also reviewed a set of issues in economic modeling. The studies did pretty nice jobs in terms of the kinds of models they used and how they generated coefficients. They did what reasonable people could be expected to do given the small amounts of resources available to them.

That was comforting from the viewpoint of trying to find something wrong. We didn't find very much. But what we did find was what Rosenberg referred to as the "dumb farmer" mentality. Although I object to that characterization because I know, from growing up on an Iowa farm, that not even the dumbest farmer would agree with the assumption that the climate is going to change significantly but nothing else will change. The reality is that if you're concerned with food security, climate is only one of many things that will change over the next half century. One of the things that we found very troubling was a lack of concern with other resources, with population change, or with the many other changes that will impinge on food security. We should be looking at a world that is somewhat more like the world that will actively exist in the middle of the next century.

The third set of things we looked at were the outcomes of the analysis. All nineteen studies measured the impact on agricultural production. But when we went into some of the other measures, such as environmental impacts, regional shifts in production, and agricultural profitability and employment, we found very limited coverage and, sometimes, inconsistent coverage. I recall one study that looked at environmental impacts using a very micro-type focus -- modeling the impact on one hectare -- and then went on to look at the regional economic impacts using large-scale regional models. It may be not too bad to use different models for different questions, but some of the assumptions driving the two were not consistent.

The most important deficiency, in our minds, was that food stocks were just not talked about except for one USDA study that looked at international trade. If food security is, in fact, what the policy maker is concerned with the possibility should be addressed that the world may become a very unpleasant place to live over the next 30-40 years as a result of massive levels of hunger. But it was not addressed in the studies we reviewed.

We made three recommendations. The first dealt with the geographic scope. The studies that were done had tended to focus on the northern hemisphere mid-latitudes to the exclusion, or almost exclusion, of Asia, Australia, South America, Africa. A lot of the world out there which grows lots of food was not analyzed. The second recommendation dealt with non-climatic demand and supply factors. Even if one is studying Saskatchewan the investigator ought to ask how the changes there relate to changes in the rest of the world. There is a rest of the world out there and a global marketing system is what ties it all together. We can't predict what the world is going to be like 40 years from now. I'm not suggesting that. But I am suggesting that one might consider how changes in population, irrigation development, land use, and other important supply and demand factors will modify the climate change effects.

A third recommendation was to shift our impact modeling in order to make the results more useful as decision support systems rather than models. I have in mind decision support systems that can be used in an interactive manner.

In summary, a lot of good technical analyses have done a reasonably good job--of answering the wrong question! When you read the studies-- particularly the summaries of the studies -- the authors go into great detail to tell you why these results probably will not hold -- that things aren't going to happen this way. That is troubling. It goes back to not asking the right question when the analyses were designed.

The good news is that we're seeing some of the modeling emphasis starting to change. Some work is now starting to look at transient climate change and the process by which climate change can be incorporated into social and economic models.

I continue to be concerned with the assertion that climate change is the problem. The implication seems to be that we need to solve it. Let's not worry too much about modeling social and economic interactions because they will be huge, whatever they are. Therefore, let's just worry about reducing emissions. All we need to do is to cut back on CO_2. If the rich developed nations will do what they should be doing anyway, not using so much fossil fuels and adjusting their life styles, the problem will be solved. We need to think more carefully about a scenario in which that does not happen. Even if there are reductions in CO_2, the rich countries will fight to maintain their lifestyles and will do so in ways that may not be very socially desirable. If the cost of cutting emissions is increasing the likelihood of World War III, do we want to pay that cost? I think those are the kinds of issues that policy makers have to deal with. And those massive dislocations in the near term aren't just simple economic issues. They're very complex political, social, and economic issues that we don't know how to even think about intelligently.

Let me turn back to the implications of the studies I have reviewed. If you look at the studies critically and just ask, "What are they saying in terms of impact?" the answer almost unanimously is that climate change on the magnitude expected over the next 50 years is not a problem for agriculture. It is just not a problem for production agriculture. I disagree with that conclusion, not because of the way the studies were done, but because I don't think they're asking the relevant questions.

Ruttan: I take it that there are two dimensions to the transition problem. One is instability. The second is the fact that it's dynamic in the sense of you're not moving to an equilibrium. It may be a disequilibrium world out there.

Sonka: There are two dimensions. One is that of heightened instability. There seems to be a consensus that global warming will result in greater climatic instability. Instability is hard on agricultural institutions. The marketing system, the credit system, and the government institutions just don't handle instability very well.

Waggoner: Do you mean variability from year to year? Is that what you mean by instability?

Sonka: Yes.

Rawlins: But that's never incorporated into the scenarios.

Sonka: You mentioned a second aspect of the dynamics. Realistically we don't know what the equilibrium is toward which we are making a transition. The people involved in policy are going to be making decisions, not about transition to equilibrium, but about the direction of change.

Rosenberg: Even on the subject of equilibrium, there's no reason to assume that it will stabilize at the CO_2 equivalent doubling or at any other particular level.

Chen: You alluded to the fact that some of the nineteen studies were very limited in resources. Do you have a sense of the order of magnitude, in money or man years, spent in those climate impact studies? Did it include the Climate Impact Assessment Program (CIAP) study, which is probably the largest, that was conducted in connection with the SST (Supersonic Transport) debate?

Sonka: No, it did not include CIAP.

Chen: Do you have a sense of the order of magnitude of the rest of the studies? I know how much was spent on Martin Parry's study at IIASA (the International Institute for Applied Systems Analysis) because I helped to set it up. In two and one-half years, they spent a total of probably $250,000 -- not including some of the contributed time of the individual researchers and individual members of the project. The money that went into it was trivial in comparison to the amount spent on some of the big international meetings that have been held in the last year or two.

Sonka: If a million dollars were spent on the studies I reviewed, I would be amazed.

Rosenberg: I would think that EPA surely spent half a million for their studies.

Waggoner: But you're considering only the agricultural impact studies?

Sonka: And only the social and economic impacts on agriculture.

Chen: It just makes a point that I'll try to make again later. Very little actually has been done in terms of actual in-depth research. This leaves a lot of room for hand waving so that people can say, "Oh, there are huge negative impacts." And you turn around and someone else is saying, "Oh, the benefits are wonderful." This just reflects the vacuum of in-depth studies that go beyond the "back-of-the-envelope" types of analysis.

Ruttan: How would you characterize the level of resources that have gone into the physical studies as compared to the studies that deal with social and economic impacts?

Chen: Several orders of magnitude in difference. It depends on where you draw the line. But even if you look only at climate models, the resources have been large. I was just at a meeting where Bob Dickinson (of the National Center for Atmospheric Research) was saying, "There are only 40 people who really know the models. We don't have many resources and we don't have our own Cray to run our models." But you're still talking tens of millions of dollars per year. And that's just the climate models. I think there's easily several orders of magnitude difference. If you look at the global change research budget that the administration talks about, it is in the $500 million range.

Rosenberg: Munson raised a very interesting point about the complexity of nature. He said that sunshine and other factors need to be considered. Most of the models used for impact studies have not considered enough of these factors. Generally, temperature is increased, say three degrees, and stays that way day and night, day after day, week after week. But climate change involves more than temperature. For example, less precipitation means fewer clouds; fewer clouds mean more sunshine; the humidity of the air changes, the windiness changes. The GCMs do not provide reliable information about these phenomena.

I want to point out part of the reason for an apparent "putdown" of impact research. Many impact studies have not made good use of available scientific knowledge. For example, in the early impact studies, modelers equated evapotranspiration with a change in temperature. If temperature goes up, evapotranspiration goes up. This is correct, of course, unless other factors such as sunshine, windiness, and humidity also change. Plants will either be bigger or smaller, depending on what the climate change does to them. The effect of CO_2 on stomatal regulation must also be considered. When all of these factors are considered a much broader range of possible evapotranspiration outcomes becomes evident. We are trying to incorporate knowledge of these phenomena in the impact studies that we're now doing.

Ruttan: Both economic modeling and physical modeling suffer from a common problem. The problem that strikes me often in our computable general equilibrium models is that they're very resistent to the introduction of new knowledge: partly because of mathematical convenience and partly because the modelers don't know that much about what new knowledge is available. I get a sense, when I hear people talking about the physical modeling, that you have some of those same problems -- that we have a lot more micro information than we're able to incorporate in the models.

Rosenberg: Yes. That's a problem, but one that time and patience can help. When studies are commissioned in a hurry, say, to meet a Congressional mandate, models must be taken off the shelf because there is no time to do anything else. But we have good information on the CO_2 direct effects and a rich agrometeorological literature is available to people who want to construct better models. The permutations can reach six orders of magnitude. The trick is to select a limited range of plausible changes.

PART THREE

Local and Regional Resources and Environmental Changes

10

Environmental Change in Eastern Europe

Zbigniew Bochniarz

Ruttan: Zbignew Bochniarz is concerned with environmental policy in Eastern Europe and the USSR, a part of the world where rather severe environmental changes are underway.

Bochniarz: I speak about environmental change in Eastern Europe and the USSR from the perspective of an economist. This is not always easy for me, because I grew up as an economist within the "monoculture" of a Marxist economy that has not dealt well with such issues as economic efficiency or the environment. There are a lot of common misunderstandings about socialist countries. Sometimes, these countries are described as planned economies. However, the evidence based on plan fulfillment contradicts this common assumption. It is true that they have plans and that they spend a lot of time on planning activities, but if you consider the implementation of plans and their usually poor fulfillment, you might better call them "planning" than "planned" economies.

In one aspect at least, these plans were very successful; that is in reaching the level of industrialization of the Western developed market economies. Unfortunately, this is not the level of industrialization as measured by wealth but, rather, by pollution. In terms of pollution per capita, they are the leaders among the developed countries. Let me give some examples.

According to a recent study by the Battelle Institute (Chandler 1990), energy related carbon emissions from Eastern Europe and the USSR (called also CMEA or COMECON countries) accounted for 26 percent of the global emissions in 1988. The contribution of the OECD countries was about 49 percent. Thirty-five years ago the planning economies contributed only 18 percent while the OECD countries accounted for 71 percent. After the "oil shock" the share of the OECD countries dropped to about 57 percent, but the share of the CMEA countries rose to 24 percent. In other words, the data show a growing contribution of COMECON to global climate change.

What are the major reasons for such developments? I would divide all reasons into two groups: objective and subjective (or man made). In the objective group I should specify two major factors contributing to growing emissions of carbon. The first is the natural resource base. Most of these countries based their industries on coal (hard, soft, or lignite), of which they have relatively a lot but which is less efficient than oil. The second is the spatial structure of the majority of the CMEA economies, which usually requires the long distance transport of energy sources for industrial activities.

The second and even more important reason is related to the Stalinist model of industrialization with its priority on heavy industry. There has been underpricing of natural resources, energy, and capital goods as well as a lack of actual incentives for economic efficiency, at both the macro and microeconomic levels.

One result of the Stalinist model is inefficient energy production and consumption. According to the Warsaw University study (Krawczyk 1987), the CMEA energy intensity per GDP is presently 2.7 times higher than in Western Europe. In terms of particulate matter pollution (dust and fly ash), the average pollution per $1000 of GDP was, for East European countries, 12.7 times that in the EEC countries. Some cynics in Eastern Europe asked why should we be concerned about global warming since we are increasing the dust and, in this way, decreasing sunshine!

In terms of SO_2 (sulfur dioxide) and NO_2 (nitrogen oxides), the difference between Eastern and Western Europe is not so dramatic. But it is still 2.5 times higher per $1000 of GDP in Eastern Europe.

The pollution problems are not just a function of industrialization, but, most important, of the very wasteful economic system which was established along with the authoritarian political system imposed after World War II. This path of development has resulted in three types of crises: ecological, economic, and political. All need to be solved in order to put these countries on a path of sustainable development.

Some data illustrate the seriousness of ecological crises. In Poland we have officially recognized 27 areas called "areas of high ecological risk" (environmental hazard) which are inhabited by about 35 percent of the population. Five of these areas are classified officially as ecological disasters. Owing to pollution and degradation of the environment, morbidity and mortality rates are significantly higher in these regions than in the rest of the country (in the case of respiratory diseases and lung cancer, about 30 percent higher). In addition to these areas we have about 50 cities that are environmentally substandard. These include about 50 percent of the Polish population. Fortunately, we have in Poland quite good environmental statistics with full access for everyone. This is not the case of the most CMEA countries, where either statistics are poor (Rumania) or the environmental data are classified (East Germany). According to dissident sources in East Germany and Czechoslovakia, the share of population living in environmentally hazardous (substandard) areas is about 60-70 percent. In the European part of the Soviet Union it is about 50 percent. When interpreting these data, one should keep in mind that the environmental standards are not, in general, as strict in Eastern Europe as in Western Europe, USA, or Japan.

There is a clear interaction or feedback among several environmental problems. One of them is acid rain or, more appropriately, acidification. This is closely related to energy policy and strategies of industrialization and hence, with global warming. Despite the several schools of thought about the metabolism of forest decline (at least six according to the World Resources Institute), none of the top experts neglects the impact of air pollution and acid rain on this phenomenon. This is a very serious issue in North and Central Europe, in the eastern part of North America, and is currently emerging in the southern part of China.

In general, acid rain is associated with emissions of SO_2 and NO_2 (sulfur and nitrogen oxides) and is part of a larger problem of acidification. Acid rain is only one of its major manifestations. Acidic compounds also can be deposited in snow, fog, and dew. They also can fall as dry particulates. The dry deposition usually takes place near the emission sources. Despite the fact that this phenomenon is not recognized as a global problem, the results of these emissions can occur up to a few thousand kilometers from the emission sources (wet deposition). According to the WRI about two-thirds of total atmospheric acidity is due to sulfur and about one-third, to nitrogen.

Acid deposition in streams and lakes and in soil and forest, as well as in buildings and technical infrastructure, causes serious damage to ecosystem, national economies, and human health. Agriculture, fishery, and forestry are among the first victims of acid deposition. Corrosion of metal structures and the dissolution of buildings and historical monuments are other examples of the effects. Let me give some data related to forest damage from a recent study of the UN Economic Commission for Europe. About 35 percent of the European forest area has been damaged. Estimates for individual countries are Czechoslovakia (71 percent); Greece and United Kingdom (64 percent); West Germany and Estonia (USSR) (52 percent); and Norway, Denmark, Poland, and Netherlands (50-48 percent).

Lake acidification is also very serious. According to the UN Report there are growing numbers of strongly acidified lakes in Canada, Sweden, Finland, and USA. In Norway, fish were depleted completely on about 13,000 square kilometers.

About 70-80 percent of Central and North European farmland is significantly affected by acidification. A meaningful increase in average soil acidification over the last 20-30 years was noted between 0.8-1.5 pH in this region. Soil acidification has had serious consequences. It has caused crop yields to decline in heavily affected areas. An even more serious effect is the release of heavy metals deposited in soil due to its acidification. Released in this way, heavy metals such as lead, cadmium, or mercury are moving either to plants or to groundwater and coming into the human food chain.

Soil and water acidification depends on its buffering capacity. It is, in turn, related to the composition of bedrock and surrounding vegetation, forest-soil, hydrology, and land use. In Central Europe the buffering capacity is much higher than in Scandinavian countries. For that reason they are more vulnerable to acidification than Central European countries. In order to increase buffering capacity and to decrease acidification in lakes, forest, and soil, several technologies have been implemented. The most popular is liming. It is, however, a very costly technology. In the case of Poland, it costs about $5 million per year.

More and more visible deterioration of the environment, as well as health effects of pollution, has led to the emergence of independent environmental organizations in Eastern Europe and in the USSR since the early 1980s. They are bringing environmental issues to the public, lobbying for effective environmental protection policies, and educating their societies. The oldest organization in the CMEA is the Polish Ecological Club; I had the honor to represent it at the United Nations annual conference of Non-governmental Organizations on Sustainable Development last September. An interesting question was asked during this conference: How would you rank global warming among the major environmental threats? Only a few representatives from developing countries, including Eastern Europe, ranked it first. It is not viewed as a major issue in that region despite the fact that the scientific commit-tees are very concerned. In these countries, problems such as acid rain, water, air pollution, and local environmental issues are of more concern.

The dramatic political changes going on this year throughout Eastern Europe, together with deteriorating economies, has produced complete distrust of any kind of planning. There is somewhat of an overreaction to past failures and the emergence of belief that only the market can save us -- that the market is good for everything. I do not share this opinion. It is obvious that we do not need the kind of planning we inherited from the Soviet Union. We need to rely on market as a principal form of regulation of our national economy. It will change the behavior of the principal actors in the market. However, I believe that we cannot leave environmental policy entirely to the market. We need state regulations -- environmental standards and strict enforcement systems. We also need some kind of incentive that is based on environment planning in the area of market failure.

We also need new institutions to respond to the global environmen-tal challenge. East European countries are far behind in comparison with their Western counterparts in sound institutional design for the environment. For this reason, last September, I organized an interna-tional workshop in Poland on market mechanisms for environmental protection. It was attended by representatives from most East European countries as well as by top experts from the United States and Japan. We prepared policy recommendations on how to use market-based incentives to protect the environment and to promote sustainable development. There is, however, a basic precondition for this instru-ment. Solutions cannot be proposed for emission trading within a "bubble." We are considering introduction of about 100 bubbles in Poland over the next couple of years. Similar approaches are discussed now in the Baltic Republics of the USSR.

Research in Eastern Europe is mostly focused on the scientific aspects of global warming. The social sciences are still far behind. I shall organize an international workshop in Poland on institutional design for the global environmental challenge next fall. This workshop will bring together leading environmentalists from both the East and West. So far, about 20 people are working on this project. [Ed. note: The workshop was held in Bialowieza, Poland, in September 1990. Representatives of six countries participated.]

We in Eastern Europe need more collaborative research with the West to improve our technologies, especially in the energy industry. We would like to reach the Japanese or Swedish standards of energy efficiency. Opening the economies of Eastern Europe has created new conditions for technology transfer, joint ventures, and even direct investment. It should help to change the structure of our economies. There is an urgent need of collaboration in institutional design at a country as well as a regional or European level. There is a lot of transboundary pollution. The solutions will require new transnational institutions. The role of the planetary economy in global warming and ozone depletion should make us more aware of the need for successful collaboration. These global problems cannot be resolved without involvement of Eastern Europe and the USSR. It is in our joint interest to develop that kind of effective collaboration.

Ruttan: Is it possible that eastern European experience may be a model for what we are likely to see in many parts of the developing world over the next several decades?

Bochniarz: If you value industrialization over the environment, you end up where we are now. This is the point I tried to make at the United Nations conference.

Crosson: Did you say that the bubble concept is actually being implemented?

Bochniarz: We are just beginning. We organized a conference and made recommendations to the government. Now there is a government team working on it in Poland. Probably next year we will have the first simple bubbles. We are also thinking about expansion of the bubble concept. In the Baltic Republics of Estonia, Lithuania, and Latvia, there is also movement in that direction. The people who established the first bubbles in the United States are involved.

Rawlins: What is the bubble concept? I'm not familiar with the term.

Bochniarz: The bubble concept is one form of emission trading. The EPA sets an ambient standard for a certain area as a ceiling. This area creates in this way a kind of bubble. All polluters within the area (under the "bubble") have to meet the standard cutting their emissions. For some, this reduction of emissions is cheaper than for others. They also can reduce emissions more than required by the standard. In this way they can earn "emission credits" which, in turn, they can sell on the market or save for their own future needs. In order to use this concept, we in Eastern Europe must first introduce and respect ambient standards and appropriate individual emission permits. On this base we can further develop emission trading.

Crosson: Are levels of air pollution in some parts of Poland now having measurable effects on public health?

Bochniarz: In the southern parts of Poland measurable effects on public health were observed. These include higher rates of respiratory diseases, lead poisoning, lung cancer, circulatory diseases, miscarriages, infant mortality, and others.

Crosson: Is there any evidence of the effects on economic productivity in these countries?

Bochniarz: There are several estimates. In general the most conservative is that we are losing about 10 percent of our GDP per year from pollution and environmental degradation. Some studies suggest as high as 20 percent. But the most conservative and documented losses were about 10 percent.

Ruttan: What about a measure simply like number of lost days of work per year?

Bochniarz: We made some estimates in the middle of the 1980s. Compared with the 1970s, air pollution and absence from work caused by respiratory diseases increased two-fold in the 1980s. For this reason we concluded that there was a strong relation between those two phenomena. In real terms, we expressed those losses as 28,561,000 days or 188,503,000 working hours lost due to respiratory diseases caused by pollution. That amounts to 4.7 days per person employed in state enterprises.

Sonka: If significant moneys come from western Europe for investments in East Germany and Poland, for example, will the investment and associated economic activity increase the pollution or will the efficiency of the new plants offset the effect of increased economic activity?

Bochniarz: A very good question! You probably remember that during last spring we conducted Roundtable talks between the opposition, led by Solidarity, and the Government. One small table organized at the Roundtable was devoted to environmental problems. Both sides reached agreement in all but one problem -- nuclear energy -- which is very promising. Both sides agreed to use a concept of sustainable development (in Polish eco-development) as the concept for harmonizing environmental and economic goals. This agreement stressed the need to increase dramatically economic efficiency as a way of reducing pollution as well as to achieve better utilization of natural resources. The marketization of the Polish economy will help to achieve this restructuring which will lead to closing many inefficient and polluting plants. Since the beginning of the 1980s we have closed several plants owing first of all, to a new independent (for the first time in Eastern Europe) Polish Ecological Club and a very militant environmental movement. The best known case was closing the country's largest aluminum plant in Skawina, which contributed about 50 percent of the total production. In reference to the topic of our consultations, I would like to mention that the Club also prepared a concept of "Ecological Agriculture" -- an organic farm belt in the buffer zone for the National Parks in 1983. In 1984 a second independent environmental organization was established in Hungary (Danube Circle). Since the Chernobyl disaster and proclamation of Gorbachev's "glasnost," environmental organizations have been mushrooming around the USSR and Central Europe.

Ruttan: From a historical perspective, we are now in the third wave of environmental concern since World War II. The first wave was very much a concern with materials adequacy. You will recall the President's Water Resources and Materials Resources reports. The scarcity issue was largely resolved by a technological response. The second wave of concern -- the environmental crisis of about 20 years ago -- was largely about micro environmental spillover effects. The prescription that came out of that was to internalize externalities. It was so cheap to pour residuals into the environment that we used up whatever cheap space was available. In this third wave things are much more transnational. It's probably more transnational in the parts of the world where countries are small than where countries are large. But I haven't seen much discussion of the institutional innovations that will be required to deal with these transnational issues. For any one national unit, except perhaps the very largest, and even, perhaps, for them, there will continue to be a tremendous temptation to free ride. I don't see very much discussion about how to design the international regimes that are going to have to be put in place to achieve some compatibility between national and global interests.

11

Temperate Region Soil Erosion

Pierre Crosson

Crosson: To start thinking about the erosion problem, it's essential to distinguish between the problem in the United States and the problems in developing countries. We know quite a bit about how much erosion is occurring in the United States, particularly sheet and rill erosion: we now have three different sets of estimates of its long-term yield effects. The estimates for wind erosion are less secure. Despite the literature, it wasn't until 1977 that we had any reasonably reliable estimates of how much erosion was actually occurring on non-federal land in the United States. In 1982, a second set of more comprehensive estimates, obtained from a much broader sampling base, became available. The two surveys--in 1977 and 1982--constitute the base of data that people in the United States use to assess how much erosion is occurring and its productive effects.

There are three studies of the productivity effects. One was done by a set of soil scientists here at the University of Minnesota under the direction of Larson (Pierce, Dowdy, Larson, and Graham 1984). They developed a Productivity Index (the PI model) that looks at the effects of erosion on certain critical characteristics of a soil. As erosion occurs, those characteristics, which are most critical to crop yields, tend to change on most soils in ways that are adverse to yield. When it's run over various periods of time, typically fifty and a hundred years, the model shows that on some 98 million acres of cropland in the corn belt where average rates of erosion are in the neighborhood of nine tons per acre per year (well in excess of what the SCS says is the maximum tolerable amount), 100 years of erosion at present rates would reduce yields by about 4 percent.

Another major study was the EPIC model developed by USDA people at ARS, and by economists at the USDA soils facility at Temple, Texas (Crosson 1986). EPIC has been run for crop-producing areas across the entire country, not just the midwest. The EPIC results show that, at current rates of erosion, at the end of 100 years cropland yields would be about 3 percent less than they otherwise would be.

The third study was done at RFF (Crosson and Rosenberg 1989). It was an entirely different approach. We did a regression analysis of the relation between intercounty differences in crop yields (the dependent variable), and erosion, topsoil depth, and a number of other independent variables, for several hundred counties in the corn belt and the northern plains. We found, over a 100-year period, no statistically significant effect of erosion on wheat yields; but corn yields would be reduced about 5 percent and soybean yields, 10 percent.

The significance of these three studies to me is that although they come at the issue in quite different ways, they all show that the hundred year effect of present rates of cropland erosion on crop yields is small-- on the order of, say, 3 to 10 percent. That's very small compared to the expected impact of technological change on yields over the next hundred years. Production or productivity losses on that order would, in effect, be lost in the noise.

Ruttan: It would be equivalent to about three years of normal productivity growth?

Crosson: Yes. I have made some estimates taking the work that I did and also some work that USDA people did with EPIC at Temple and estimated the economic costs of long-term loss of crop yields in the United States. The effects, as you might expect, are very small.

As far as the United States is concerned, the long-term effects of soil erosion on the capacity to produce agricultural output is trivial at present rates. A much more significant erosion-related problem is the off-site or off-farm damages. These include the effects on water quality, particularly of sediment after it leaves farmers' fields. These turn out to be at least on an order of magnitude greater than the economic costs of lost productivity. In research on erosion in the United States, much more attention ought to be given to water quality issues relative to the amount of attention being given to productivity issues.

It is not surprising that on-farm damages should be much less than the off-site damages. Ruttan made some reference a little while ago to the externalization of costs. The loss of soil productivity is an internal cost for farmers. The farmer bears that cost and, therefore, has an incentive to do something about it where it significantly affects either his income or the present value of his land. He has no such incentive to deal with off-site damages. Those are external as far as he is concerned. They are internal, of course, to the total society. And the fact that those external damages are almost surely greater than is socially optimal is a reflection of institutional failure.

Earlier I distinguished between the erosion problem in the United States and that in the developing countries. The contrast in our knowledge is startling. Despite the fact that we read statements about how much erosion is occurring in the developing countries and what the productivity consequences are, the fact is that nobody really knows. A recent article in the *Journal of Soil and Water Conservation* (Colacicco, Osborne, and Alt 1989) by one of the leading people in this area, emphasized that we really don't have a clear idea of even how much erosion is occurring in developing countries, let alone what the productivity consequences might be.

Some recent work at the World Bank carefully examined in Indonesia this twin issue of on-site productivity losses attributable to erosion and off-site damages. The preliminary results suggest that the on-site costs are, in fact, higher than the off-site costs. Much more of this kind of research is needed in places all around the world where there is reason to believe that erosion is a significant problem. There are areas in the Himalayas, in the mountainous parts of Latin America, and in East Africa where a lot of anecdotal information suggests that not only is erosion high but it also may be having long-term consequences for productivity. I'm not arguing that the situation in the United States is typical of what is happening elsewhere in the world. The point I'm making is that we don't know enough about what's happening elsewhere in the world with respect to soil erosion to be able to make well-grounded statements about the importance of the problem and what we ought to do about it. My hunch is that when such studies are finally done, they likely will show that in most areas the off-site damages are likely to be higher than those on-site damages. It is important to consider that we need to know much more than we do at present about the erosion problem in the developing countries in order to be able to address issues relating to the sustainability and enhancement of agricultural production capacity.

In a special issue of *Scientific American*, Rosenberg and I had an article (Crosson and Rosenberg 1989) suggesting strategies for long-term agricultural development. The focus was the emerging pressures on the natural resource base in response to rising demand for food and fiber over the next several decades -- well into the next century. The question we asked was, what can the community of concerned people do to accommodate the increases in demand and relieve the pressures on the resource base? We particularly focused on land resources, water resources, and biological diversity. We concluded that there will have to be a continuation of new technologies that permit farmers to respond to the rising demand for food and fiber, in a manner that is consistent with environmental and other social costs. Some of the institutions are now in place for developing the appropriate technologies and are relatively well established -- the national agricultural research institutions and CGIAR system. The more difficult issue is likely to be the development of the institutions that would correctly signal the emerging stress on environmental resources. The development of the institutions needed to reflect the social scarcity of those resources would be more difficult to develop, primarily because of the difficulties of establishing property rights in water and in biological diversity. As a consequence, we concluded that the institutional challenge, or the challenge to develop appropriate institutions, would be more difficult than meeting the challenge of developing the appropriate technologies.

I was also impressed, as I read through this special issue of *Scientific American*, how much weight was given in the other chapters to the issue of institutional design. The design of institutions for managing our resources in ways that give proper weight to the various social values we place on those resources is emerging as a key element in our thinking about development processes and, in particular, about issues of sustainable development, not only in agriculture but, also, in the total economy.

Allen: Each time I go to Africa, I get depressed about the difficulty of bringing about the institutional changes needed to generate the appropriate technology, getting that technology transferred to the farmer, and creating the right incentives. I hope that my interpretation of what I see is totally off base. I'm curious whether you gave any attention to bringing about the needed institutional changes in some of these countries where population is growing at an explosive rate and the land resource base is degrading.

Crosson: My impression is that Africa is a worst case. But there has been some rather impressive progress in other parts of the developing world. As you know, Africa emerged from colonialism later than these other areas. There was systematic discrimination against agriculture pricing policies, investment policies, and rural education and infrastructure.

Ruttan: My sense is that it's not going to be easy to generate rapid growth in agricultural production in Africa. In the European and Asian cultural systems, there is a sufficient similarity in property rights and family structure to give us a good deal of intuitive understanding of how institutions work. My sense is that for both property rights and family structure, which are closely related to each other, the system in much of Africa is so different it appears obscure to outsiders. Furthermore, the family structure was often deliberately obscured by local people during the period of colonial development.

Rayner: I would like to disagree here. It's not so much that the workings of the traditional African land tenure and kinship systems are obscured. In the ethnographic literature, you can find very good descriptions of them. I think the problem is largely that our institutional arrangements are not directly transferrable. Nor has the expertise on how those traditional land tenure and kinship systems work been transferred from the academic world of anthropologists to, or accepted by, the development community. Nowhere is this clearer than in the land tenure issue. Economists have gone in with western ethnocentric assumptions, such as, "If you want to get people to plant trees you have to give them tenure of the land." In some of those systems they have well developed tenure rights to the standing crops. I see that more as our institutional failure. It's not that the information doesn't exist in the West to understand if there is a willingness to use it. It's much more the dominance of Western neoclassical economics that obscures our understanding of institutional behavior.

Sanchez: Let's backtrack to the question on how extensive the erosion problems are in the tropics. Erosion is much more severe in the semi-arid regions, the areas where the land is bare during the dry season and gets torrential rains at the beginning of the rainy season. This includes most of the semi-arid regions of Africa as well as some highlands areas. Parts of Ecuador, in the Andes, are tumbling down visibly and grossly. Erosion is less of a problem where there is ground cover continuously throughout the year, as in much of the humid tropics. That doesn't mean there's no erosion. But it's certainly less of an issue.

We need to think in terms of having a ground cover throughout the year. It doesn't matter whether it's weeds or tropical forest.

Rawlins: Crosson, have you looked at the positive impacts of sediment delivered to off-site locations? Many major river systems have very productive deltas that are the product of upland erosion. King, in his book, *Farmers of Forty Centuries* (1911), discussed the husbanding of sediment as a means of enhancing productivity in the orient. I wonder if anyone has taken a systematic approach to assessing the productivity of whole river basins and balanced the decrease in productivity in eroding highland soils, where the climate may be less desirable, with the increase in productivity in deltas and river valleys.

Crosson: As far as I know, it hasn't been done on a river basin scale. There was a little work done with the EPIC model to examine the effects of deposition on productivity on different types of landscape. They found that it makes considerable difference when you take deposition into account. But the model simply assumed that increases in soil depths because of deposition would have had positive effects. That is not necessarily correct if soil depth is not limiting. Larson may know more.

Larson: I know about that study. But let me first comment on your earlier remarks. The 3 to 10 percent loss in productivity due to erosion just doesn't tell the whole story. For about 25 percent of the 400 million acres of cropland -- about a hundred million acres -- the losses would be on the order of 10 to 15 percent. And if you take the 10 percent most erosive land, the losses would be in excess of 25 percent. Now, it's that 10 percent that we have to concentrate on. If we lose 25 or more percent of productivity, that land's going to go out of crop production. And it should! The conservation reserve, and to some extent the cross compliance, should be aimed at the land where soil losses are in the 10-15 percent or above range.
While I agree with you that the NRI estimates of 1977 and 1982 are the best data on erosion we've ever had, they still leave a lot to be desired. As you know, the Universal Soil Loss Equation (USLE) gives us point estimates. They don't take into account the landscape effect. They don't take account of deposition. A lot of things are not included. They estimate only sheet and rill erosion, not gully erosion. There's also a need for more careful definition of what is acceptable or tolerable and what isn't. Much of what we accept is not much more than folklore. It's

107

not based on actual measurements. It's a committee decision that was made 40 years ago. I was probably part of that committee.

Cross compliance is going to become a very big issue. As you know, by 1991 or 1992, farmers who aren't complying with these acceptable limits, whatever they are, won't be eligible for price support or deficiency payments. A recent study looked at the same farms that had been used in the NRI estimates and now were certified for compliance, and with little or no change in practices the two estimated amounts of erosion were far different. A lot depends on what goes into the soil-loss equation and on the erosion trends that we define as acceptable.

I recently reviewed a paper written by Harold Dregne (1990) in which he reviewed all of what he considered quantifiable data on erosion from Africa. He came up with only a very few small studies. Many of those were for very limited physiographic areas, so the conclusion is that we don't know. You can see erosion everywhere in Africa, but there's inadequate quantitative data.

Crosson: The numbers I gave on erosion induced productivity losses in the United States are national average productivity losses. They are meaningful numbers when the question is how the nation's capacity to produce will be affected by erosion over the long term. I have argued that the USDA programs ought to be focusing on areas where there's reason to believe that those losses might be high; but where farmers may be unaware that over the long term, they will be losing significant amounts of productivity. Some of the work that Larson did indicates that the relation between productivity loss and top soil loss is linear over a long stretch of top soil losses and then turns down sharply. Farmers are not likely to know that. They can be moving along having erosion of 20 tons per acre per year, and suffering little or no productivity loss when, in fact, they may be nearing the edge of a cliff. They may have no reason to know that they are approaching a threshold. They judge what the future may be by what their past experience has been. The USDA would be well advised in its soil erosion productivity research to focus on those soils where it's believed the relation is nonlinear.

Chen: One minor point: there are estimates that some of the flooding problems in the lower Mississippi Delta are precisely because erosion has been reduced. There is a benefit from sedimentation in terms of keeping existing deltas from suffering from local sea-level rise. Larson just related that 10 percent of the land suffers from a 25 percent yield decrease due to erosion. Is that 10 percent the most productive land or the most marginal land?

Larson: The most marginal.

Chen: So, in fact, the proportional contribution to total agricultural production may be a lot less than that?

Larson: Yes. When you look at a piece of land in terms of the effect of erosion, you've got to look at both its inherent productivity and the rate of damage. The USDA has tended to look only at the amount of erosion. Some very deep soils can erode for a long time without much damage. But it you have only 20 inches of soil over bedrock, you can't erode very long before you're in trouble. We've argued that you want to look at both of those things here in Minnesota in our RIM program. The USDA hasn't seen fit to follow our lead.

Rawlins: We hope the new methodology being developed to replace the universal soil loss equation (USLE) will not be used simply to calculate average annual soil loss from erosion, and that the databases being put together to support it will help to move in this direction. The Water Erosion Prediction Program (WEPP) should make it possible to make risk assessments, taking into account weather scenarios, soil, and management factors at specific sites. If WEPP is used simply to replace the USLE in calculating annual average soil loss we will have failed to achieve our objective.

Clark: You've just indicated that, for the United States, there are three different studies that give us a handle on the impacts of long-term erosion on productivity. Is there written down someplace what the minimal program, in terms of time and resource requirements, would be required to get comparably credible estimates for a useful sampling of situations around the world? I encounter vast piles of material and new sets of equations coming out of FAO, UNEP, and even USDA. I have been involved in extensive efforts to push the international geosphere/ biosphere program in the direction of incorporating a very large coordinated measurement effort that is directly relevant to the ability of the world system to support people. But I have yet to see an outline of the minimal requirements. I believe that it is the sort of thing that the IGBP program would be extremely interested in having because they are being beaten up all around the world for arguing self-indulgent pure science, with no connection to food security, agriculture, human health or anything else. They're not willing to leap all the way out into policy research. Hard physical measurements guided by theory is great.

Ruttan: What did it take to generate the data that was used to analyze the effects of erosion on crop productivity?

Larson: To start with, it would require a generalized soil map of, say, Africa, along with a topographic map and some kind of a weather data base.

Crosson: There are two steps. One is the collection of reasonably accurate erosion data and the second is using those data in models that tell you something about the effects on productivity. Both were very expensive in the United States then but perhaps less so now. With the EPIC model, the USDA can now cover the entire crop production area of the country.

Ruttan: But doesn't that require a lot of sites where you're actually collecting data?

Crosson: Data were collected in 1982 for close to a million points. The 1987 NRI was much less intensive -- about 300,000 data points -- and it came to essentially the same conclusions as the much larger 1982 study of the productivity effects of erosion.

Larson: You could probably come up with a first approximation with much less than 10,000 points. But, it would have to be done by soil scientists or engineers. You would have to have the collaboration of someone who really knew the soil and the landscapes in the country. You can't do it in Washington or St. Paul. It could be done at a reasonable cost but with not quite the accuracy of the U.S. analysis.

Clark: A small, quick and dirty working group that would flesh out the proposal you just made would be very timely. Half the countries of the world are already flying under the sustainable development banner.

Ruttan: What about the big project at the University of Hawaii? Doesn't it have a very large data base that is relevant to this issue?

Sanchez: It's not an erosion data base, it's a crops simulation data base. FAO has done a map of soil degradation in Africa. I think it's terrible.

Larson: My experience in Africa is limited, but erosion is everywhere. You can see it and recognize the seriousness of the problem. But it must be quantified to get any credibility. A program like the one Clark proposed is needed. We could argue about the scale. But some precision is needed to give it credibility. I was shocked when I read the article by Harold Dregne, who is a respected soil scientist. He could come up with only about 8 or 10 examples of reasonably adequate quantitative measurements of erosion in all of Africa.

Bochniarz: We have been told many times that the U.S. Soil Conservation Service program could be used as a model for other countries. What about the U.S. soil conservation program in terms of its impact on soil erosion and soil quality? What is the net result of that program? Is it a model for other countries? Are there adequate data on desertification in developing countries? What is the relation between soil erosion and desertification? To what extent does soil erosion contribute to desertification?

Larson: Let me respond to the question about the value of our research rather than the SCS program. A half dozen different groups from different countries have tried to extend and use our Minnesota model. Actually, the EPIC model and the work at Temple is even better. I think the answer to your first question is that the research could be transferred. I don't know much about desertification. I think that's even more of a no-man's land in terms of reliable numbers.

Crosson: There are two ways of looking at the question on whether the U.S. Soil Conservation Service could be a model. As far as I know, there haven't been any definitive estimates of how much difference the Soil Conservation Service has made over the last 50 years in reducing erosion and protecting soil productivity. It can hardly be anything but positive, however. There is less erosion and less productivity loss than you otherwise would expect, but the question of net economic benefit is less clear.

The second part of the response is that the effects could have been much better. There has been a lot of waste of Soil Conservation effort, not because the SCS was indifferent or wanted to do things that way, but they were under constant Congressional pressure to put projects into areas where there was no significant erosion threat. I think it has been a positive difference, but not nearly as much as it could have been.

Cheng: I want to share a little observation that I made in the People's Republic of China. Even in the United States, much of soil conservation policy is inconsistent with the economics of farming. In China, for many years, a great effort was going into reforestation to prevent soil erosion. But during the last few years, after the household incentive system was introduced, farmers suddenly found that they don't have the time to continue the conservation practices. Erosion problems have worsened. China conducted a nationwide soil survey and trained people down to the township level. Tens of thousands of villagers and technicians were collecting soil samples and doing analysis. This was the Second General Soil Survey which began in 1983. But up to now, there hasn't been a single soil map generated from those data. And the data that were collected were not very useful for soil management. They are too generalized. So it's not going to be very easy, as Sanchez pointed out, to get a generalized map of Africa or a country of comparable size like China. The effort in this country has taken many, many years.

Rawlins: I think we need to follow up on Clark's proposal. The databases are being constructed now. Agriculture needs to have an input into the global change program to answer some of the questions being raised. We have teams developing predictive technology for water and wind erosion composed of representatives of the USDA and Interior agencies. These teams are developing cooperative agreements with groups in Brazil, Canada, Australia, and Israel to validate these models under their conditions. We would be more than happy to extend this network by cooperating with others to make certain the appropriate data are collected to answer the important questions.

Herdt: I just want to make an observation about the need for additional research or information gathering. Governments in the Third World, especially in Africa, are under great pressure. I'm not sure how many of you saw *60 Minutes* last night. A lot of the shots were from Tanzania. But the program started off with the news that 40,000 children die every day in the Third World from a combination of disease, malnutrition, lack of water, and many other things. Governments in the Third World are not insensitive to disasters of that magnitude, but they are incapable of addressing them. They're incapable of preventing them. That's one illustration of the kinds of immediate pressures that these governments face. Many of you have worked in the Third World and you know that in every field there are immediate needs that push back research, especially research that has a payoff in the far distant future. In soils-related research, there is a great scarcity of human resources. Cheng mentioned the training of thousands of people in China. The data were collected, but you can't find the results -- there has been no pay-off.

Clark: But we also know that one of the surest ways to increase the number of trained people is to build training into an apprenticeship mode around solving very particular problems that bridge the line between immediate relevance and basic science. It is because this issue has a certain amount of substance to it, much more so frankly than in many of the areas I see my natural science colleagues trying to cultivate through developing country collaboration, that I push it a little harder than I might otherwise. It's irresponsible to place demands on that incredibly scarce resource without designing the effort to expand the capacity to sustain the food resource base.

12

Tropical Region Soils Management

Pedro A. Sanchez

Ruttan: We shift now from the temperate region to the tropics. Pedro Sanchez has spent his entire career trying to understand how to sustain and enhance agricultural production on tropical soils.

Sanchez: I want to talk about the humid tropics -- to focus on tropical deforestation and possible solutions. One possibility of abatement of global climate change is to reduce tropical deforestation. A lot of technology is sufficiently developed to be worthwhile testing. This issue is not only related to global change but, also, to other important issues, such as the preservation of biodiversity. Reducing tropical deforestation is relevant to global change and to many other issues. The process of deforestation is very complex. Most of it is population and economics driven.

In the Third World, the driving forces are population growth, limited fertility, and land tenure inequities. The consequence is a landless rural population that is faced with three choices. One is to stagnate in areas of high demographic concentrations, such as the Andean region, northeast Brazil, Java, and others. A second is to migrate to the urban centers. In the case of Latin America, we have the dubious honor of having the largest cities in the world already. A third option is migration into the humid tropics. In the Andean highlands and in northeast Brazil, agriculture is being pushed into steeper and steeper areas which causes a lot of erosion, siltation of reservoirs, and smaller and smaller minifundia every generation. Those who migrate burden the urban carrying capacity. This is evident in the large cities in Latin America. When I lived in Lima, a city of six million people, half of them had no sewage services, electricity, or water.

The ones who migrate to the humid tropics, either spontaneously or sponsored by government colonization projects, end up in a world they do not understand. The peasant farmers are not familiar with the shifting cultivation systems used by the indigenous inhabitants. They practice shifting cultivation in a nonsustainable manner. The forest fallow periods are shortened and the productivity of the system declines. In some countries, particularly Brazil, a large part of the deforestation is caused by land speculation. Tax breaks and credit incentives from the governments induce large landowners or companies to clear land, partly for cattle production but mainly for land speculation. Currently, new laws are beginning to slow things down. In either case, the result is an unsustainable agriculture that results in economic failure. Most people who migrate to the cities thus exacerbate unemployment. Traditional societies are disrupted. The result is a cycle of further deforestation, soil and land degradation, loss in genetic diversity and an accelerated greenhouse effect.

What can be done? The data show that 80 percent of the tropical deforestation is caused by nontraditional shifting cultivation -- by small farmers who clear and burn a couple of hectares of land a year, mainly to grow food. The other 20 percent is caused by land speculation, logging, urban development, road construction, and other unit works. But the vast majority of land clearing in the humid tropics is by small farms. Are there alternatives? Many people believe that we can stop deforestation by making it illegal, by outlawing shifting cultivation. Shifting cultivation is illegal in many countries -- Indonesia, for example -- but that doesn't stop it. Other policies include producing food outside the humid tropics to reduce the pressure on the humid tropics. Consider, in Brazil, the now very productive Cerrado region south of the Amazon. Brazil should grow its food there rather than in the Amazon. But people are still migrating into the humid tropics. Sometimes migration is even supported by the government, as in the transmigration programs in Indonesia. Other people believe that you should put a fence, so to speak, around the tropical forests, make them into huge national parks, and not touch anything. That is not realistic. The national parks have to be defended. It is not possible to stop people who want to clear forests to grow food.

Rawlins: But shifting cultivation does not leave the abandoned land in a permanently cleared state. Don't the clearings grow up to forests when the cultivators leave?

Sanchez: Traditional shifting of cultivation, with low population densities, allows forest to grow back over 20 or 30 years. Even so, the forest loses much of its genetic diversity during that transition because when it's cut, the second growth is not so diverse. It has also lost a lot of the carbon because the secondary forest will never grow to the size of the primary forest. But it would grow back and have no long-term detrimental effects on soil quality or soil erosion. The trouble is that, with high population pressures, farmers cannot afford to wait even 15 or 20 years to reclear. When we got to the area where I worked in Peru, in the early 1970s, the fallow period was about 20 years. Now the fallow period is about two to three years. In its pristine stage, shifting cultivation is fine from the ecological point of view, but the cultivator remains in perpetual poverty. Other alternatives are needed. What are they? Are alternative technologies viable?

There are a lot of myths or misconceptions about the agronomic potential of the humid tropics. The soils are low in fertility, but the statement that it isn't possible to grow crops continuously in those soils is incorrect. It is not correct that clearing a piece of forest will turn the soil into laterite, or that the soil will be so compacted it is useless for agriculture. This may occur in about 3 percent of the soils, but certainly not 97 percent. It is not correct that most of the nutrient cycling that goes on between the forest and the soil bypasses the soil. Those are wrong assertions. What has been found by research is that sustainable agriculture is feasible. The classic form of high-input agriculture that feeds the world, including the green revolution areas in the tropics, is technically feasible in many areas in the humid tropics. It may not always be logistically or economically feasible because the roads may not be in place and the marketing infrastructure may be absent.

Therefore, a menu of options is needed and has been developed. I would like to share with you a very simple landscape model. It first shows the heterogeneity of the humid tropical landscape. There are beaches, low flood plains, and high flood plains. There are areas that flood every year or perhaps once every 10 years. There are high terraces with low fertility soils, but flat and relatively easy to cultivate technically. There are also low fertility soils in hilly areas that are difficult to mechanize. Finally, there are mountains, usually with young soils. A series of options has been developed for each area that is very sustainable. The most obvious one is to grow paddy rice, as in southeast Asia, in the high flood plains, in areas where water can be diverted from the river or low lift pumps can be used. Rice is consumed in all these countries. In Peru, there has been great progress in paddy rice production in the Amazon, based upon very simple technologies transferred from Asia, combined with plant breeding to adapt the varieties of local soil, pests, and pathogens and to local tastes.

For the high flood plains that do not flood regularly, or for the high terraces that are flat and easily mechanizable, continuous cropping with lime and fertilizer inputs is feasible. But this requires a good road infrastructure, accessibility to market, and a reliable credit system. In general, the physical and institutional infrastructure is underdeveloped, except in areas around the large cities of the humid tropics. Indeed, there are some large cities: Belem has about two million people, Manaos is over a million people, and Iquitos is over a half a million people. Although this proposed system is technically feasible, it is economically viable only where the infrastructure exists. For most areas, it is not economically feasible. In such areas, a low-input cropping system is probably the best alternative. The idea is, instead of changing the soils to meet the plant requirements, to change the plants to produce under conditions of soil acidity and low levels of fertilizer. We have been able to design such systems by finding varieties of key crops, rice and grain legumes, that are productive at low pH levels -- that are perfectly happy at pH4 or so, which is very acid. By designing an improved slash and burn system, we have been able to grow about seven crops of rice and cowpeas in rotation over about three years without any fertilizer or lime. And we had pretty decent yields. This has been done by trying to capture and recycle the nutrients. We capture the nutrients from the slash and burn secondary forest, religiously using all the crop residues. This recycles most of the potassium. After a while, the fertility of the soils does go down, simply because we are taking a lot of nutrients out, such as phosphorus, in the form of crop harvest. There's no way we can totally put that back in the form of crop residues. As fertility goes down, weeds increase. It is then time to abandon the field. But instead of letting it grow to a secondary forest fallow, we use acid tolerant legume fallows, such as kudzu, which cover the ground rapidly. Kudzu smothers the weeds and after a year or so you can slash and burn it and start the circle again. But some fertilizer will be necessary to replace what has been lost by crop removal.

The next stage is a transition to high-input agriculture if, in the meantime, the physical and institutional infrastructure has been developed. Otherwise the transition may be to agroforestry systems. The agroforestry systems are based on the use of trees that are acid tolerant and well adapted to the region but can produce food or timber or some other marketable product. There are several that are very promising. You probably have not heard of most of them. But this is part of the beauty of the biodiversity that exists in tropical forests. Let me give one example.

The peach palm (*Bractis gasipals*) is very well adapted to acid soil conditions. Within four years it starts producing fruits that look and taste like sweet potato. The nice part is that it can produce about 10 tons of fruit per hectare per year over a period of 15 or 20 years. You don't have to plow. It's edible. The local people like it. It's also a good feed for monogastrics, such as swine and chickens. It could replace corn in feeding systems. This palm is multipurpose. You can cut off the apical stem and market it here and in Europe as a gourmet food, as heart of palm. This particular palm regrows after its stem is cut, so you don't destroy the tree as in some other species that produce heart of palm. Other fast-growing trees can be used to produce charcoal or wood. Many are legumes, so they fix nitrogen.

Pastures have a bad reputation in the Amazon. They deserve it because they're poorly managed pastures that have been introduced mainly from temperate Brazil. The producers there use no fertilization, employ poor cattle management, and do not incorporate legumes in the pastures. But well-managed pastures that use acid tolerant grass and legume species, which were developed mainly by the International Center for Tropical Agricultural (CIAT) in Colombia, have proven to be very productive.

So there are a number of options. Groups of researchers involved in these studies believe that we're probably at the same stage as the Asian green revolution was in the late 1960s. The prototype technology is there. A combination of these alternatives, when applied to areas that are undergoing intensive deforestation by people who want to settle and produce food, perhaps slow the deforestation, given the right policies. It's not going to happen unless the government policies support these more sustainable systems and discourage shifting cultivation and other forms of destructive land use.

We calculated how many hectares can be saved from deforestation for every hectare put into these different management options. In the case of flooded rice, farmers are able to grow about 11 tons per hectare per year. That's in two crops averaging about five and a half tons per hectare per crop. Under shifting cultivation, upland rice yields are one ton; you need to cut 11 hectares to produce 11 tons of rice. For every hectare in flooded rice, you could save about 11 hectares from additional deforestation. The ratio in low-input cropping systems is less. But for every hectare that's put into even the low-input, sustainable systems, you might be able to save from 5 to 10 hectares of forest every year.

There are viable alternatives to shifting cultivation. Farmers don't shift and cut forest because it's fun. Cutting a tropical forest is extremely hard work. It also means moving households or commuting by foot farther and farther from their fields. There may be options that could slow deforestation if policies are developed to encourage sustainable systems. About eight countries account for over 80 percent of the world's humid tropical deforestation. A program that concentrated on those eight countries would cover most of the problem areas. I shall not speculate how much this would affect global warming. In the last year I've seen reports that tropical deforestation may account for as little as 10 or as much as 25 percent of global warming. The point is that something can be done now.

Crosson: Tell us a little bit more about the policy changes that would be necessary in order for these systems to be adopted on a wide scale.

Sanchez: Some are quite simple. The Agrarian Bank in Peru should pay the farmers for their rice at the time it is sold, rather than four months afterwards. With an inflation rate of 40 percent per month, farmers are discouraged from growing rice if they can't get paid for the product when it is marketed. Other changes are more complex. The transport infrastructure has to be developed to reduce the cost of getting inputs, such as fertilizer, into the communities and getting the products out. Also, there should be some disincentives against further land clearing.

Crosson: You mentioned that the absence of adequate infrastructure is a primary impediment to adoption of high-input systems. Would you expect to see both the high- and low-input systems expand if they provide the necessary infrastructure?

Sanchez: It depends on individual country or regional situations. A country like Brazil that can produce plenty of food in the South probably should not provide incentives for high-input systems in their Amazon. It would be more appropriate to focus on some low-input agroforestry systems. A country like Peru that is deficient in basic food grains may want to encourage rice production. The point is that there is a wide range of options, depending on the local ecological and economic environment.

Rawlins: I have seen claims that the long-term harvest of nuts and other natural products from the natural forest would produce a higher return in the long run than could be obtained by opening the land for pasture or crop production. Are there other economically viable but less destructive ways of using the forest?

Sanchez: Very few. Natural rubber is one. But while it may preserve the forest, it provides no way for the rubber tappers to escape from poverty. If you want farmers or gatherers to stay where they are, that's one way to do it. Selective cutting has been very disruptive. In order to select and harvest the few marketable trees per hectare in humid tropical forests, you destroy a good deal in the process. Forest enrichment, in which you cut swatches two or three meters wide in the forest and plant improved species, has not worked. There's too much competition from the natural vegetation. Pastures under forests have been tried. The pastures are poor and the forests are partly destroyed. It may be better to leave the forest as it is. If you want to save the forest, don't touch it. Do something else around it. From a forestry point of view, a tropical forest is a mess. It's a horrible tangle of hundreds of different species. A production forester does not want to mess around with a humid tropical forest. He would rather have a nice, well-planted, well-spaced uniform stand of trees. There are a few exceptions, like rattan, which is a vine that can be used for baskets and furniture. There are things like the rosy periwinkle, a little flower discovered in Madagascar, that's used as a base for chemotherapy. But, you can't use a tropical forest and keep it as it is.

Chen: I read something about using iguanas and other small animals for food. Could enough of them be raised to get a reasonable income from the meat?

Sanchez: Iguanas are from the semi-arid areas. But there are also a lot of small rodents in the rain forests that are delicious. If you're going to produce them commercially, it will probably mean raising them outside the forest. I think most animal scientists are not optimistic about the possibility of using new animal species on a large scale. At the local level, however, they have a place.

Rawlins: What about biomass power generation? Species mix would not be an obstacle.

Sanchez: No. That could be another option. You could have income from the forest and replace it with fast-growing biomass. *Inga* species produces over 15 tons of dry matter per hectare per year during its first three or four years. The oil palm, as I mentioned yesterday, produces more oil per hectare per year than any other crop. You can put it directly into diesel engines. There are other species of palm that might be used for oil production in the Amazon. We may be able to harvest the tropical forest for biomass, but I hope we will not. Biomass can be grown better elsewhere. Also, keep in mind that the carbon accumulation in tropical forest may be as large as 200 tons of carbon per hectare. Every time we clear a tropical forest, we're releasing large quantities of CO_2.

Waggoner: Did you say 200 tons of carbon per hectare in the standing crop per year?

Sanchez: The range is probably between 50 and 250 tons of carbon per hectare. When the standing crop of a virgin tropical forest there is in equilibrium, it's just sitting there. When it is cut and burned, most of it either goes right into the atmosphere or decomposes and goes later. It cannot be recaptured. Some soil organic carbon also decomposes after land clearing.

Rawlins: If you put the forest biomass through a power generator to replace fossil fuel, you haven't lost it.

Waggoner: Then presumably it could be replaced with fast-growing trees that would have a positive accumulation of carbon per year.

Davis: What he's saying is that it would be very hard to regenerate to that very high original level.

Sanchez: It could take 50 to a 100 years.

Rayner: Another problem, if you're looking at this from an energy perspective, is where to locate the generator. You're not going to locate it near a forest that will not be replacing. For biomass energy to be economically feasible, it must be based on high-intensity biomass production. There's no other way to do it economically. That means a closely planted, fast-growing species.

Davis: Can the production systems be used on present pastures? Can you reclaim those present pastures?

Sanchez: Yes. There are two or three ways. One is to improve the present pastures by planting some of the improved species, especially legumes, in strips. In small pastures, you go in and spread a little bit of rock phosphate and plant legumes. They gradually take over and improve the pasture. When the pasture has been degraded badly, you can burn the degraded pasture then fertilize and seed it. It depends on the nature of the degraded pastures. A lot of pastures are degraded because they're infested with grasses that animals don't use. Others have been compacted or degraded physically and chemically. They require a different type of renovation than ones that are just degraded in terms of weeds.

Davis: What are the economic and policy issues there? Who owns these degraded pastures? Who controls them? Who uses them? What happens to them?

Sanchez: Well, it varies from country to country. Most of it is owned by individuals. The large-scale pastures are often owned by absentee landlords. They often don't manage them well because they're just sitting there speculating on the land price. The farmers are in the 50-100 hectare range, as opposed to thousands of hectares in some parts of Brazil. These small units are interested in improving their incomes. You can work with them.

Davis: But in Brazil, where they've cleared forest and replaced it with pastures, there's no way that a small farmer actually has the capital to go in and reclaim a pasture, is there?

Sanchez: The small farmers in Brazil have pastures. In Rondonia, where they may have 50 to 100 hectares, they are very interested in making those pastures become more productive. There are low-cost technologies that can be used to gradually reclaim the pastures. This is in contrast to what happened in eastern Brazil along the Belem-Brasilia Highway where so much deforestation took place about 10 years ago.

Huge pastures were developed in that area. It's probably going to take some government intervention to make those people change.

Allen: You didn't say anything about fish culture or animal culture, beyond ruminants on pastures. Are there any other animal systems that would be useful if integrated into the total systems you talked about?

Sanchez: There are other possibilities: fish like tilapia. Instead of having two crops of rice a year, you could have one rice crop and nine months or so of fish culture. There is also some potential for the water buffalo in this system. The water buffalo eats low-quality pastures that no self-respecting cow would touch. It is the ideal draft animal in the humid tropics. They have great potential in the humid tropics of South America.

Ruttan: Realistically, isn't the problem that, until the infrastructure comes in, you can't get the stuff in and you can't get the stuff out without hauling it on your shoulders? But even when the infrastructure is developed and the more intensive system becomes economically viable, it will always be a poor system. The producers are not going to get rich, even if they work as hard as a Taiwanese peasant. It seems to me that, given the pressures of people against the resource (and that's not going to stop), you're going to see continued opening up and destruction of the forest. I find it very difficult to believe that the governments of Brazil, Peru, Venezuela or central Africa are going to be able to stop that process, given their limited administrative capacity and their political structures. As you know, even the Philippine government cannot keep peasants from invading the national park on the mountain right behind the International Rice Research Institute.

Rawlins: Reducing population pressure must ultimately be the means to reduce the exploitation of tropical forests.

Clark: It's too late.

Crosson: If the systems you are working on are economically more attractive to the small farmers than the slash and burn system, why don't they adopt this system rather than slash and burn?

Ruttan: In the areas I've looked at, infrastructure is a major barrier. Just ask yourself: "How many kilometers do I have to carry a bag of fertilizer or a bag of rice before its value becomes zero?" Not very far. There must be dense infrastructure to sustain intensive agriculture.

Rayner: Surely in a place like Rondonia, Ruttan is right. A decade and a half ago there were only four or five paved roads in the whole state. Now, if you look at the aerial maps, at least two-thirds of the state is covered by a very regular grid pattern of paved roads. The pattern of deforestation has followed the developments of that road system very closely.

Sanchez: The area in Rondonia affected by roads is still very small; no more than 5 percent of the state. But you are right. Where the roads are is where people are. They get a land grant of about a hundred hectares. Then the question is, what do they do after they clear and cultivate for a couple of years?

13

Soil Fertility

Robert D. Munson

Munson: I have had a long-term interest in soil fertility. I started out working on nitrogen soil tests at Iowa State. At that time, we were working on nitrogen soil tests and trying to figure out how to predict how much nitrogen you needed to apply for crops. In the 1960s nitrogen became so reasonable that we proceeded to forget what we had learned. During the mid-1980s, we came back to some of the same issues because of environmental concerns. We are now making nitrate-nitrogen soil tests and coming up with estimates of the nitrogen rates to apply to improve crop yields, increase use efficiency, and decrease environmental impacts.

An individual can do a great deal with a pencil and calculator. You do not always need computers to solve soil fertility problems. Several years ago a dealer asked me, as a consultant, to develop a prediction equation to determine the amount of fertilizer nitrogen to apply on his customers' fields. After reviewing available data from experiments in Illinois, Iowa, Minnesota, and Wisconsin, I developed an equation for testing. We tested the value of the "program" for a specific customer. We had soil test information as well as spring nitrate-nitrogen tests on samples taken to a three-foot depth. The dealer asked, "How much nitrogen do I need to apply to grow 150 bushel corn?" For that yield and level of measured nitrate-N, the answer was, clearly none. For that site, the dealer's hand-harvested yields for six hybrids averaged 162 bushels per acre while those that were combine-harvested averaged 157.8 bushels. My estimate for the amount of nitrogen required to obtain 175 bushels only missed the mark by three bushels, and gave a net return of 33 cents per dollar invested in nitrogen. Plant leaf analysis from this corn indicated that both potassium and zinc were probably limiting the recovery and use of the soil and fertilizer nitrogen. My point is that we currently know a great deal more than we are effectively using in making wise fertilizer recommendations.

The other point I would like to make is, we do not get something for nothing. Each crop has an "internal" nutrient requirement or efficiency for each essential element to produce a unit of yield, i.e., per bushel or ton. For most crops, aside from carbon and oxygen (from carbon dioxide) from the air and hydrogen (from water), thirteen other elements usually come from the soil. The crop's internal requirement may vary somewhat with maturity, variety, or hybrid, but, for example, for corn, it is relatively constant over a range of yields from 80 bushels up to over 300 bushels per acre. Also, when a crop is harvested and "leaves" the farm, various amounts of those essential nutrients also leave.

In general, the other elements or nutrients with which we are usually concerned for crops include nitrogen, phosphorus, potassium, magnesium, and sulfur. But in many areas of the world we also need to be concerned with the micronutrients, sometimes called trace elements, because of the small amounts present; they include boron, chlorine (chloride), copper, iron, manganese, molybdenum, and zinc.

If one considers soybeans, relative to their internal nutrient requirement, it is a different value than for corn but, in general, the same principle holds. [While soybeans, being a legume, have the capacity to symbiotically fix a portion of their own nitrogen, much of that capability is dependent upon the availability of other essential elements, including cobalt and nickel, with beneficial effects for selenium.] A similar phenomenon is true for wheat. Phosphorus tends to be more variable than nitrogen and potassium. This is interesting because of the role that phosphorus plays in photosynthesis and energy exchanges within crops. One would expect it to show the least variability, but there is evidence that once high-yield levels are reached, crop yield per unit of uptake variability may be least for potassium.

Soil tests can be used with a fair degree of accuracy to predict the need for fertilizer phosphorus applications. Also, soil phosphorus levels can be increased through applications of phosphate fertilizers and/or manure. While test levels decrease over time with cropping and very high phosphorus soil tests, additional fertilizer phosphorus will not increase yields until the test levels are drawn down below the "critical" level for the given soil or group of soils. In some cases for intermediate yield levels it may take up to ten years to reach that "level." What is difficult to predict is how a crop variety or hybrid, when combined with different cultural and management practices, may interact with the applications of row or starter phosphate on high-testing soils in different regions. For example, with earlier planting of spring small grains or corn in northern areas, responses may be apparent that would not exist on soils with similar tests in warmer areas. Also, effects of reduced tillage or no-till may decrease soil temperatures, increasing crop responsiveness to banded phosphate and/or potash. Furthermore, some corn hybrids may be much more responsive to either phosphate or potash than are others grown on the same soil and site. From an environmental standpoint, early evidence indicated that mixing applied phosphate into the soil through tillage or subsurface banding decreased phosphorus losses due to runoff or erosion. Crop responses to given rates of application were also usually superior with banding or knifing-in.

The yield or output of a crop grown on soils with different soil tests for a given element produces different efficiencies. Also, when a given amount of fertilizer nutrient is applied, it produces different yields on different soils, depending upon the crop, soil test level, soil-fertilizer reaction products produced, and their subsequent availability and uptake by the crop, as well as the seasonal weather. Also, soil microorganisms are usually competing with the crop for some elements. Often, the recovery of fertilizer phosphate is considered to be about 20 percent, while crop recovery of nitrogen and potassium applications are considered to be around 50 percent. Of course, if soil tests or application rates are excessively high, recovery drops off dramatically. Just because a nutrient is not recovered the first season following applications, however, does not mean that it is "lost" or that it causes environmental problems. That is particularly true for phosphate and potash applications. Environmentally, there is far more concern with soil and fertilizer nitrogen.

In thinking about soil and fertilizer nitrogen, we need to keep in mind the nitrogen cycle. It should be recalled that in looking at N recovery, unless ^{15}N (isotopic N) is used, we obtain recovery by differencing the N released and recovered by the crop from soil organic matter, with that taken up by the crop from the same pool, plus that from the fertilizer N. We need to remember that on the control plots, where no N is applied, soil organic matter must be expended to provide the N for that yield. For example, in a southwestern Minnesota corn study, the 24-year average yield of the control plots was 69 bu/a. Over the period yields ranged from 11 to 141 bushels, depending upon the rainfall and season. The internal nutrient efficiency or requirement to produce a bushel of corn usually averaged about 1.12 pounds of N per bushel. Therefore, in this example, an average of over 77 pounds of N would have been taken up by the aerial portion of the crop each year.

But, if N recovery was 50 percent, about 154 pounds of N would have been needed in the system to grow the crop (with lower yields, recoveries of 70 percent are achievable). If the harvested grain removed 0.72 pounds of N per bushel from the field, the net annual N loss from the field would have been about 50 pounds per acre per year. For soils with 5 percent organic matter, that would mean a net loss of about 1000 pounds of soil organic matter per acre the first year and an increasing amount in subsequent years as levels decrease with time. Therefore, the net loss would be well over 12 tons per acre for the 24-year period. For this soil, about a third of the organic matter would have been lost.

On the above soils profitable yield increases were obtained from fertilizer N application. The yields of the control plots in general are not too different from those obtained in experiments in Ohio, Illinois, or Iowa. The main difference appears to be that as one moves from east to west, the level of nitrogen that can be used effectively decreases, because water becomes limiting. It should further be said that for corn, with a harvest index of roughly 50 percent, for every bushel increase in yield, an equal amount of dry matter will be left in the field to potentially add to the soil organic matter if C:N:P:S ratios are maintained. Therefore, depending upon the initial level of soil organic matter, the judicious use of nitrogen and other fertilizer nutrients over the last forty years actually improved the productivity of many soils. One only has to study trends in the long-term data from the historic Morrow Soil Fertility Plots at the University of Illinois to observe the benefits of improved soil fertility on yields and productivity. The Rothamsted Experiments in the UK, perhaps the oldest known soil fertility plots in the world, provide further testimony on the use of liming materials and fertilizers to maintain and improve soil fertility and soil productivity.

In nitrogen balance field studies the nitrogen recovered amounts to between 70 to 75 percent. In studies on rice, however, recoveries in the crop and soil can be in the high eighties, even to 90 percent when proper timing and sources are used. (N recovery by the grain in some studies may account for 50 percent of the N applied.) My early view, based on available research data, was that most of the nitrogen not found had been lost to the air and not to groundwater. But we know that even under the best conditions on control plots, there is going to be some nitrate-N that gets to tile lines and, perhaps ultimately, to groundwater. The measurements that have been made in southern Minnesota on unfertilized plots show that the water coming out of the tile line contains about 13 part per million (ppm) nitrate-nitrogen. When 100 pounds of N per acre is applied, the yield increases to about

145 bushels, but the tile water contains 41 ppm nitrate-N. My point is that you cannot produce crops without having nitrate-nitrogen in the soil solution. If people think that we can grow legumes -- for instance, continuous soybeans, or alfalfa as a source of N -- and not have nitrate-nitrogen present, and some being lost to tile lines and groundwater, they are just kidding themselves.

Dr. Gyles Randall put it this way: "There is a cost of doing business in crop production," and having nitrate-nitrogen in the soil environment is part of the cost. As soon as we tilled our soils, especially those formed under tall prairie grasses, and speeded up the decomposition of soil organic matter, we started to release nitrate-nitrogen to the environment. (In the early days, because of low yields and poorer technology, we probably lost more N to the environment than we do today.) No matter what we do, we are going to release some nitrate into the environment. The other thing to remember is that the nitrogen in the system is rate driven. It becomes very important to choose the right rate, timing, and cultural and management practices in order to grow higher yields so that as little nitrate-N as possible is left in the system at the end of the season. Other things to remember are to use soil testing and profitable rates of other essential nutrients, with land-use management that will optimize the efficiency of applied nitrogen and the moisture available.

14

Information Systems for Soil Management

William E. Larson and H. H. Cheng

Larson: My thesis is that many of our environmental and natural resources problems could be eliminated or minimized if we used better geographic information systems. We have all seen cases of bad land use in agriculture, forestry, rangeland, urban development, waste disposal sites, and others. Today, of course, we pretty well understand the geology and the macro geographies. But we still don't recognize and take into account the micro spatial variabilities in soils and landscapes. The micro spatial variability is usually captured in our soil surveys. The National Cooperative Soil Survey has the goal of mapping the cropland in the entire nation by 1992. That's a tremendous job. I don't know what it costs nationally, but in Minnesota, the federal and state governments spend about $4 million a year on soil surveys.

In the past, the county soil surveys were usually reported in a thick document with 50 - 100 pages of maps. Most people just wouldn't use them because they were too complicated and cumbersome. But the modern maps have been digitalized and put on computers. A farmer in Minnesota now can walk into our County Extension Offices and within 30 seconds he can have the map of his farm up on the screen. A specialist can tell him about the characteristics of his land and how to interpret the soil survey. In Minnesota, we now have about a third of our counties digitalized. Most other states aren't that far along. Some are a bit further. It costs money but it makes the soil survey a usable tool. We must also combine the soil surveys with good geographic information systems. We in soils have made the soil surveys but we have not always done the landscape mapping in enough detail to make them as useful as they might be.

Geographic Information Systems (GIS) consist of geographical referenced data (soil, topography, vegetation, hydrologic) as well as the necessary hardware and software. With the proper input, we can display all the features that are important in management of a landscape and we can determine their interrelations. For example, using GIS one can determine erosion amounts; route sediment movement and deposition as well as water storage and movement; determine crop yield potentials; calculate soil organic carbon sequestered; display plant nutrient amounts and needs; and estimate other important management tools in a landscape setting. It allows us, for the first time, to put together a quantitative total land use and management scheme that takes into account all features of a landscape. Up to now we have usually generalized from a field-by-field basis. I believe the GIS is a giant step forward in a total natural resource management system.

Let me give an example of what can be done. Soils vary a great deal in their chemical and physical characteristics. Even a seemingly uniform landscape often has great differences. The public only sees the surface. There's a lot of variation below the surface. We took a 50-acre field in southwest Minnesota that is fairly typical of the glaciated areas of the upper midwest. The published soil survey had seven mapping units in that 50-acre field. Surface texture varied from loam to clay loam. The slopes varied from about 0 to 10 percent. Our published crop equivalent rating indicated that the potential for corn yields on the seven mapping units varied from 65 to 145 bushels of corn per acre. Any soil scientist knows that you have to match fertilizer inputs to the potential production of that soil. It would seem ridiculous to fertilize that field in a uniform manner. That 65 bushel soil does not have the same response to nutrients as the 145 bushel soil. Our published fertilizer recommendations vary on those different mapping units from 40 to 150 pounds of nitrogen per acre per year on corn.

Equipment has been developed that will apply differential rates of fertilizer as it goes across the field. The equipment carries straight nitrogen, straight phosphorus, straight potassium, and herbicides in separate bins. It mixes them on the go and applies different rates as it passes over the field. These rates can be based on the soil survey, on soil tests, or on historical yield records. The computer facilities are in the cab of the machine to vary application rates as the machine goes down the field. These machines are commercially available.

If you fertilize the 50-acre field at the 150 pounds of nitrogen rate, which is appropriate for the best soil and which is probably about what a farmer would do, then only about 10 percent of the field would be fertilized at the correct rate. But 77 percent of that field would get 30 pounds more than would be recommended, 7 percent would get 50 pounds more per acre, and 7 percent would get 110 pounds per acre more. Likewise, these seven mapping units have a hydraulic conductivity that varies by about five-fold. The absorption coefficient for the two pesticides vary about five-fold. This means that on those seven mapping units, some of the herbicide recommendations would be only half what they might be on others.

On this 150 acre field, using the machine that I described, you could save about $8 an acre. At the same time, potential leaching of the fertilizer into the ground waters would be reduced. I calculated that you would save about 1,500 pounds of nitrogen on the 50-acre field. And if you assumed that all of that was going to leach into ground waters, then it would account for about 10 parts per million of nitrate in the top foot of the ground water. That may not be perfectly accurate but it does give you an order of magnitude. I also calculate, using our model, that if erosion continued at present rates on that field, at the end of one hundred years some soils would be unchanged and others would experience about a 12 percent reduction in productive capacity.

My point is that we are now developing the information technology that will help us to achieve sustainable agriculture. We must do a much better job of matching the soil and the landscape characteristics with the management, including nutrient use, pesticide applications, erosion control, and others. With modern digitalized soil surveys and modern equipment, we are making progress. The data aren't available in many places. But that's our charge for the future: to develop the data bases, including soil surveys, landscape data, and weather data bases.

I'm often asked if I picked an example that's not typical. I think that field is typical. When I was in Idaho in the Paloose area a couple of weeks ago, they were talking about what they called "Catina management." It is the same thing we're talking about, but by a different name. Last weekend I was in the South Carolina coastal plain giving this same sort of talk. I asked Pat Hunt how many mapping units he had in a 60-acre field right adjacent to the station. He said 18 different mapping units on this coastal plain land that looked as flat and level as this floor. The variability I described is normal. In the future, we will benefit by taking that variability into account in our management systems. We have the technology right now. But the data bases are still underdeveloped.

Clark: Is this $8 an acre or so you thought you could save figured with current market prices of the inputs? Is that a big incentive relative to the investment you have to make? Let's assume the data base exists, so it's just the added investment to be able to use the machine. I have no feel for the scale.

Larson: It could represent about 15-20 percent of the fertilizer cost.

Waggoner: Munson started out with the example of what he did on the back of an envelope. I want to ask Cheng and Larson about all this savings that you were going to realize with these elaborate machines. Couldn't the gains be realized without this intermediary of the elaborate machine?

Larson: I told someone at the break that when I was a boy in Nebraska my father used to put me on the manure spreader and say take it out and put it on those eroded knobs. That was farming by soil variability. You can do it by hand, you can do it a lot of different ways; it doesn't have to have that elaborate machine.

Waggoner: Yes, but my point is whether the machine actually is necessary. By concentrating on the machine we could imply that the principle was only applicable to Europe and North America.

Rawlins: It is a very good point. We always need to distinguish between principles and tools. But there are places where you simply must have the machines to apply the principles. Having sufficient nitrogen in the root zone for highly productive crops does not have to result in pollution if the right management is used. In Florida some horticultural crops are grown with a plastic cover to protect the root bed from leaching. This management practice forces the water to infiltrate between the beds, leaving the nitrogen intact. We need to be smart enough to develop other management practices that decouple water and chemical transport. But application of these practices may require specialized tools.

Larson: The company that manufactures this equipment has sold about 70 all over the world.

Clark: So at the moment, if one simply conceives of the data base itself as a public good that should be publicly funded, then one doesn't even have to add that investment cost to the fertilizer and pesticide prices in order to have an incentive that is sizeable.

Larson: The cost of establishing the data base is. But there are the water quality benefits as well as cost reductions.

Crosson: Not for the farmer, though, unless his own well is being contaminated.

Larson: That's frequently what happens. Our farmers are also concerned with the environment.

Clark: I think the key is that if you can make money on the issue and feel good about it to boot rather than having to impose a cost to do good, then it's just all that much more of an incentive.

Crosson: Does the manufacturing company have a program to develop the yield potential information? It would seem that there would be a market incentive. Is there any evidence that that's happening?

Larson: A third of our counties have the data for the whole county digitalized. Farmers can get that from us or from their county. If the county doesn't have the data, the company will digitalize the farm for a fee. We charge the counties for digitalization because they use it for land tax assessments and lots of other things besides farm management. It costs about $25,000 for a county to do it. That is about $1,250 for a six mile square township -- practically nothing.

Rawlins: I have been working with some commercial companies on the possibility of developing position-sensing instrumentation for tractors and harvesters. Experimental harvesters have been developed that can measure yield continuously as they move through the field. The combination of these two technologies would make it possible to develop highly detailed yield maps. Using last year's yield as a surrogate for soil productive capacity throughout the field may be a good first approximation for varying the distribution of fertilizer automatically through computer control.

Position sensing would also provide guidance for machines to follow the same track each time through the field. This would provide the means for implementing Larson's zonal tillage concept. A compact traction zone could provide all-weather access to the field to apply chemicals or harvest in a timely manner. A specifically tilled infiltration zone could route infiltration around the root zone, decoupling water and chemical transport. Fertilizer and other chemicals would be applied only to the protected root zone, which is tilled to enhance rooting. All the components to practice zonal farming exist, at least at the experimental level. Someone needs to take the responsibility for integrating them into an economically viable system. The first customers could be experimenters who need to put out yield trials. They could design the yield trials in their computers and then program the seeder or fertilizer applicator to halve the application rate in some sites and double it in others. Then when the field is harvested the computer could automatically calculate response functions from its numerical yield map.

It seems to me that last year's yield map actually may provide a finer structure for applying variable fertilizer rates than the soil map.

Larson: Perhaps. I often like to use the analogy of a dairyman. A dairy farmer would never feed every cow the same grain ration. He feeds the cow depending upon her milk response. Similarly, we've got to feed every soil depending upon its response potential.

Munson: There are mobile soil sampling units that, through satellite geopositioning, can record exactly where a sample is taken. When the analyzed soil test results are fed into a microchip on an on-board computer, it provides the information that can be used to change the rates and combinations of plant nutrients (essential elements) that are applied as the fertilizer applicator moves across the field.

Cheng: You were talking about precision sensing. Right now it's done by laser, by triangulation, and it's not too accurate. But satellite technology is coming on stream that will let us pin point the application of fertilizer by computer within 10-foot intervals.

Davis: I recently attended a forest soils conference in which very much the same philosophy was being proposed for managing a national forest to reduce the inputs of herbicides. The USFS was trying to develop soils maps which would tell them the potential of a site and then manage the forest for what would grow there anyway so that they could reduce the cost of herbicides and other forms of management.

Ruttan: My wife would like to make sure they don't spray when she's out picking blueberries. Let's turn now to Cheng.

Cheng: Larson spoke of a number of the things that I had in mind. But I would like to pick up on a number of things discussed in the last couple of days, and clarify a few points.

Yesterday I mentioned the adaptiveness of farmers. When we were talking about "dumb farmer" models, I made the comment that the problem is not the "dumb farmers" but the dumb modelers. Even those of us that grew up on a farm often have an image of farming which is decades out of date. For example, in many of our global models, we tend to treat all crops the same. But even all corn is not the same. Some producers use as many as 10 different varieties each year.

When I was still in Washington state, we had a very well-known wheat breeding program. There was one breeder who has been quite successful in breeding winter hardiness into wheat. Yet every time I went to wheat growers' meetings, I would find the farmers using varieties from Oregon rather than from Washington. The varieties are less winter hardy but have other characteristics that are probably more desirable. They were willing to take that risk because they were betting on the weather. We had not had a really hard winter since 1968. They're willing to take these risks. They are much more adaptive than our scientists, who just keep on trying to breed winter hardiness into the wheat. We have the same problem with a plant pathologist. He was always trying to identify varieties resistant to snow mold but we haven't had snow mold for 20 years. We need to think about the adaptiveness of our science and our scientists.

Let's move on to a few other points. I'd like to clarify something about methane gas. I think we need to be careful in thinking that any time we find waterlogged or saturated soils we will automatically produce methane gas. Methane is only reduced at the lowest redox levels. There are a lot of other oxidized species in the soil which must first be reduced, particularly iron, manganese, and nitrate. De-nitrification is far more important. Even in rice fields we find, because of water movement, there is an oxidized zone in the soil. There are certain soils in the rice field that do produce methane. But we have to be very careful about these generalizations. We talked yesterday about the problem of cadmium in fertilizer. In California and Japan, cadmium is not a problem.

Of the various points that Larson and I have discussed, the one I would like to reiterate and elaborate on is the importance of understanding soil variability. We all know that there are many types of soils on a landscape. The difficulty has been to extrapolate information developed from a point on a landscape, such as from a pedon or even a field plot, to describe accurately the characteristics of the landscape or the biological responses thereon on an area or regional basis.

Let's take the problem which we have been discussing here about the potential impact of global climate changes on agricultural productivity. A key aspect that has not been discussed extensively is the effect of climate change on the soil carbon budget and terrestrial carbon cycling. We understand the carbon cycling processes well enough but we have little of a quantitative handle on how or how much soil can modify the impact of carbon dioxide in the atmosphere. Depending on the cropping systems, tillage, and residue management practices, we may put more carbon dioxide into the atmosphere from faster organic matter decomposition, or we may sequester more carbon in the organic form in the soil from greater biomass formation, both by plants and by soil microorganisms. Some of our carbon and nitrogen transformation models can help us to predict the trend and estimate the magnitude of carbon dioxide production or carbon sequestration. Using the GIS capability, which Larson mentioned earlier, to combine the soil survey information base, the climate resources base, cropping history, soil management practices, and geostatistics, we now are able to extrapolate process information developed from a point or small area on the landscape to estimate the magnitude of changes on a landscape basis.

In addition to the GIS capability to account for spatial variability, we have also developed process models to describe temporal changes in carbon cycling so that we can evaluate the potential impact of climatic changes on soil carbon budget. Furthermore, we can even estimate how changes in cropping and soil management practices could change the soil's capability to sequester carbon. For instance, certain tillage practices can increase the level of soil organic matter contents. just imagine, if we can increase the organic matter content of the soil by just 0.1%, we could sequester an additional 2,000 lb of carbon per acre of land. This could affect the carbon dioxide level in the atmosphere tremendously. What I have described means that we now have not only the capability of accounting for the spatial variation of soils and soil properties but, also, the tools to manage the soils according to their various characteristics.

I must quickly add, however, that to increase soil organic matter level even 0.1% may be easier said than done. When we start to cultivate a piece of land for row crop production, we usually note a decrease of soil organic matter level. Therefore managing land to maintain the soil productivity is a challenging task. I believe that many large-scale developing projects, such as in the tropics, would fail because they do not include a proper soil management policy.

In recent years, I've traveled quite extensively trying to understand about wet soils. They look beautiful under their natural conditions. If they are kept wet, there is good leaching capacity and they generate very nice lush growth. But as soon as they come into cultivation, as soon as they dry, a lot of their characteristics change. You have to be very careful in managing these soils. We have to recognize that each soil has to be managed differently. We may be able to reduce the material inputs by increasing our inputs of knowledge and information. This means not only data intensive, but also analysis intensive systems. The problem with many of the models is that there is no way to verify them. We've worked for years on the water quality models, large landscape watershed type models. EPA has worked for almost 20 years. Finally they gave up because there was no way to verify them because of the variability. So most of our modeling effort must become much more process oriented, rather than just regressing the apparent cause on effect.

But let me just go one point further. What we have described is only one application of the concept. In the soil survey information system, we have the data base not only for agricultural production, but for water quality. We are including data on pesticides, their leachability, and their adsorption characteristics in each soil. We also have decision making aids for wildlife habitat, for recreation purposes, for building constructions, and for a number of other purposes. There's a whole list of information being gradually built into the system to make the soil survey information useful, not just for farming, but for total management of our soil resource and for maintaining environmental quality.

Herdt: I wonder if everybody accepts your last proposition that, by going to a more closely controlled use of inputs, you're going to achieve all the social as well as all the private goals.

Cheng: The economists and the sociologists are going to have to work with us to answer that question.

Sanchez: I agree with Cheng I think he is in a win-win situation. If he can increase the efficiency of inputs and decrease pollution, that is a lot to accomplish. Isn't this what sustainable agriculture means in the U.S. context?

Clark: It is at least moving toward sustainable agriculture. I think the point is that if you can substitute information for inputs, that has very positive environmental externalities. The information is less polluting than the pesticide or fertilizer that you displace. You may still end up with too much nitrogen or too much pesticide drifting around your system. But you have made the next stage of the problem easier to address.

Larson: It's a big job to make all these surveys, digitize them, develop all the auxiliary data bases, and interpret them. Software development is a big job.

Waggoner: Now you're beginning to say that perhaps it's expensive.

Crosson: My understanding is that once the data are available, then it is relatively inexpensive to make that information and related services available to the farmer. But it's still not entirely clear to me how the cost of collecting the information in the first place should be handled. When I refer to a win-win situation, I interpret it to mean that the costs to the society of making these practices available at the farm level are less than their social benefits, and that the cost to the farmer of adapting the practices is less than the benefits he derives from them.

Chen: But there are gains to society over and above those to the farmer. Thus, society has to make the investment in order to get the returns.

Crosson: The nice thing about win-win is you don't have to do anything other than provide the farmer with information about the practices. In his own interest, he will adopt these practices, so he wins and we win. It may still be in society's interest to do something to achieve these practices, even though it would not be in the farmer's economic interest to pay the full cost.

Rayner: In the report that we just completed for DOE on policy options for private sector responses to climate change, we actually looked to the issue of smart machinery, and suggested that two things at least would be important in compressing the market penetration time. One would be some kind of public invention support program to get the right devices developed. And the other was some public expenditure on a demonstration program for the technologies.

Cheng: The Minnesota Legislative Committee on Resources has been funding our accelerated soil survey. The state recently passed a new constitutional amendment that will create an environmental trust fund; it will eventually build up to a billion-dollar base so that we'll have somewhere about $50 or $60 million every year to improve resource management.

15

Pests and Pathogens

Richard Jones

Ruttan: We've talked about many things that have implications for entomology. Climate change affects the things Jones is concerned with. And what he does about the things he's concerned with is going to affect the environment.

Jones: My charge is to discuss constraints in agriculture in the next century as they relate to pests. Although I'm an entomologist, I will try to talk about pests in general -- insects, weeds, and diseases -- because they are three significant worldwide constraints to agricultural production.

As most of you know, we suffer significant losses in agricultural production due to pests. The latest estimate is that about 13 percent are lost just to insects. When you add weeds and diseases to that, it would be considerably larger. Insects, weeds, and diseases are constraints on the types of crops we grow in certain parts of the world and even in certain parts of the United States. Without pesticides, for example, it would be almost impossible to grow potatoes in the Red River Valley or lots of other places in the United States. Cotton is another commodity that would be very difficult to grow without pesticides. These commodities are on the pesticide treadmill.

What changes do we expect? Many will relate to the costs and benefits from pest control. The farmer does a cost-benefit analysis to decide whether or not to use a pesticide. Social cost-benefit calculations enter into public policy and regulatory decisions. New issues are being added to the cost calculations. These include greater attention to food safety and water contamination. The question is how to incorporate the environmental insult from the use of pesticides in a manner that influences farmers' decisions. The farmer responds primarily in terms of perceived change in profit. This calculation differs around the world. The cost-benefit relation in Africa is different from that in the United States. The pesticides which we have depended upon heavily for the last 30 or 40 years are under increasing pressure, not only because of increased costs but, also, due to increased pest resistance. Many companies are finding their pest control business less profitable and at least one major company either is considering cutting its R&D budget or getting out of the pesticide business.

Whether we like it or not, we have to look for more alternatives in dealing with our pest problems. One alternative is biological control. Certain types of biological control can be less expensive but many will be more expensive. Another is host-plant resistance. A third is the use of biocides-pathogens that perform like an insecticide. Integrated pest management is a concept that has been around for the last 25 or 30 years. Adoption has been slow because it is information and management intensive. For example, we have developed economic thresholds and we recommend that farmers spray only when the populations reach a threshold. But farmers tend to be reluctant to make the counts. In certain parts of the country and for certain crops like cotton, consulting has become an active business. Consultants do these weekly counts. But it's not very widely adopted. The reason I mention it is that sociological considerations enter farmers' decisions. They like a simple operation.

We've talked about global warming here and how it will affect pest problems. It will affect weeds just about like other plants. We can expect increases in the range of certain weed species. In terms of insects, two or three degrees centigrade change can affect the range of a pest by several hundred miles. In the case of Minnesota, for example, if that happens, then it means that probably we'll have another half of a generation per year of European corn borer. It will increase the number of generations of an insect pest. This will increase the demand for pest control.

I don't know that anyone's really examined the effects on insects of an increased CO_2 level. At lower oxygen levels metabolism is inhibited because insects breathe by diffusion, and an oxygen-carbon dioxide ratio determines their metabolic rate. I assume that if CO_2 increases and oxygen stays the same, then you can expect metabolism to slow somewhat in insects. But this may be more than compensated for by the increased temperature.

The fertilizer effect of increased CO_2 actually will make a more nutritious plant, one that is more desirable to insects. For example, one theory about the spruce budworm outbreak is that it is nutritionally related. So you do get a fertilizer effect of CO_2, then it could have some impact on insect populations and outbreaks.

Moisture is an important factor also. It's particularly important for pathogens. Increased humidity improves the conditions for pathogen development and there would be more problems with plant diseases. On the other hand, lack of moisture creates a favorable environment for some insect pests like spider mites and grasshoppers. During the last three years in Minnesota, with drought and above average temperatures, both of these problems have increased significantly.

One thing that influences the severity of pest problems is that our world is shrinking because of modern transportation. We wind up with a lot more exotic pests; they are causing most of our problem in the United States. The problem probably will get worse instead of better. I don't know how to deal with the problem because it's not possible to effectively regulate movement of people and goods in and out of countries. However, a number of these exotic pests are good candidates for classical biological control. If you go back to the country of origin and collect the natural enemies, there's a good possibility that you can reduce the equilibrium population of the exotic pests. It has worked with a number of exotic weeds and insect pests.

Biotechnology will have an impact on pests, particularly in the area of host plant resistance. We can expect to see great progress in transferring genes that will provide resistance to plant-specific pest species or to pathogens. It's not a panacea, however, because once you put that gene in the plant a process of co-evolution of pest and host will ensue. An example is the case of wheat and Hessian fly. It's been going on for 30 years. We have to continually release new varieties of wheat to maintain resistance to the Hessian fly. That type of thing will continue to occur as biotechnology is applied, only the types of resistance will be a little bit more dramatic. Progress has been limited thus far because only a few different suitable genes have been identified, such as the BT toxins, to put into plants. So far, they aren't tissue specific, so the toxin goes all over the plant. It remains to be seen how acceptable these toxins are going to be in the food chain. That's a big issue and it hasn't been resolved yet. Given our increased concern with toxic chemicals in our food supply, it's likely to become a significant concern.

Another one of my concerns is property rights. Using this type of breeding to develop resistant plants puts the whole ball game into the hands of the seed companies. It could cause the price of seed to become more expensive: a lot more expensive if one company has a monopoly on a suitable gene. This raises the question of who will develop these materials and to whom the property rights will belong. In the past, universities have been very heavily involved in crop breeding. The role of the public sector in developing and maintaining ownership of this genetic material is becoming a significant question.

Another issue affecting the pest problem is chemophobia. That's the current public perception of the danger of chemicals. It is going to be with us for a long time. We can expect to see more attention to how synthetic insecticides or pesticides compare to natural pesticides that actually may be a lot more carcinogenic than the synthetic pesticides are. The issue of the naturally occurring carcinogens that we eat regularly will receive increased attention. This is not just a scientific issue but a question of public perception. We must be conservative in the use of our pesticides, but we need to be realistic. The public perception of the danger of some of these chemicals has little relation to reality.

How will changing agricultural practices affect pests, particularly such practices as reduced tillage? Rotations for the most part, reduce pest populations. Reduced tillage probably will result in increased pest populations, particularly insects. Fall surveys of the European corn borer in counties in western Illinois indicate over five borers per plant, which is higher than it's been since World War II. And that's in an area of Illinois where there is a lot of reduced tillage. One doesn't want to generalize from such a combination of events, but it's an indication of something that needs more investigation.

What are some alternatives to pesticides? I've mentioned some: classical biological control, biologicals that behave like insecticides, and host plant resistance. Integrated pest management will be more widely used. There's always talk about big population reduction programs, such as the use of sterile males, as in the screwworm control program. People are still thinking about how to pursue such techniques with other pests. Currently a project is going on to eradicate the boll weevil from the southeast. It's making some progress but it is working against very powerful forces of natural selection. Only a pest that is causing tremendous economic cost can justify such large- scale expensive techniques.

A few more things deserve mention. One is habitat destruction, such as deforestation. We think it's a big problem but we don't know how big because we don't know what's there in terms of either insects or plants. A big effort to inventory what is there is needed. We know a lot more about our solar system than about the biology of our planet. And this is a big deficiency. E. L. Wilson said it's more important to conduct a biological inventory of the planet than the human genome project because the genome will still be here 30 years from now but these other genes will not be. But that kind of biology is not glamorous. It involves drudgery work, like the soil survey. It's hard to get funds to do it, but until it is done, we won't even know what we're losing.

Another issue is the relations among pests, health and agricultural production in Third World countries. The battle against insect vector diseases is being lost. More people are dying of malaria today than ever before -- a million a year in Africa. DDT didn't eliminate malaria and it didn't eliminate mosquitoes. But it eliminated all the people that were working on malaria and mosquito vectors (because of the perceived lack of need). Now we have very few medical entomologists. We don't know what the impact of this million deaths a year on agriculture in Third World countries will be.

One concern is the declining support for graduate training and research in the pest-management disciplines. We need to give greater attention to urban entomology. In the present budget climate, any increase for urban problems will be at the expense of agricultural efforts. Another thought is how to use constraints, such as global warming, as an opportunity to make changes that we know need to be made. One change that will make agriculture more sustainable is diversification. In Enterprise, Alabama, there's a statue in the city square with a big boll weevil on top. The statue is there because the boll weevil mandated the diversification of agriculture in Alabama.

Abrahamson: Have you speculated at all on the insect vectors and human parasitic diseases that might affect Minnesota with a somewhat warmer climate? For example, why don't we have malaria here any more?

Jones: Mostly because we have eradicated the breeding sites of the malaria vector in the United States. We recently introduced into the country the Asian tiger mosquito. It's a vector for encephalitis. It has moved as far north as Chicago. If it gets three to four degrees warmer, it could move into Minnesota. Lyme Disease could turn out to be very significant. The Minnesota Department of Health thinks we don't need to worry about it very much, but the Wisconsin Department of Health is concerned with it. It's a tick-borne disease that, in its latter stages, can be very crippling; it can be controlled with antibiotics if caught early. With warming, we can expect to see the additional diffusion of vectors from the south into the upper midwest. For example, right now

Heliothis is a problem in sweet corn in southern Minnesota about one year out of five. If we get a few degrees warmer it could be a problem four years out of five.

Waggoner: Integrated Pest Management (IPM) is an example of more intensive use of information that has been around for a long time. What can you learn from the experience with IPM?

Jones: Pesticides are used heavily in cotton. IPM has been successful in cotton because the cost of insect control is very high -- up to $150 an acre. When we came along with additional tools, such as fall stalk destruction, host resistance, and economic threshold measurements, it was possible to make significant dollar savings. I think we see less IPM in Minnesota because the cost of insect control is lower here than it is in cotton or some other crops in the south. A soybean field in Minnesota is practically sterile, but in Missouri the soybean field is alive with insects and has to be sprayed.

Ruttan: My perception of the slow adoption of integrated pest management is that IPM represented a political compromise between agricultural entomologists and ecologically oriented entomologists. They papered over their differences with the IPM label, but the technology wasn't there, in fact.

Jones: That's partly correct. It was a concept that was ahead of its time. An important component of pest management is accurate predictive models. We're still working to develop the predictive models for a lot of crops. You have to have the data before you can feed it into the model. The data have been expensive to obtain. The hope of integrated pest management, of integrating all of the pest control practices, has fallen short because of lack of information. For example, we're now running trials in potatoes to determine the effects of water, verticilim, and leaf hoppers to try to find out what's causing something called "potato early dying." It's not due to the simple effect of any one of these, but to some combination of events. IPM will require data on the interactions between these different pests.

Rawlins: I think that's an extremely important point. It bears on your comment that we will probably be moving toward more diversification in the future, which will be more management intensive. One reason farmers have gone to monoculture is that it is less information intensive. You start diversifying, particularly when you bring animals into the picture, and you have to develop a whole new set of talents. One challenge to research is to learn how to deal with complexity and to provide the tools that will help farmers to deal with that complexity.

Clark: I don't want this five-degree temperature change number to become established by repetition as common knowledge. The current accepted view, to differ somewhat with Davis, is that if we're concerned with summer temperatures, which are the ones most significant for most of the insect and pathogen vectors in this part of the world, we should be talking about smaller increases for the 30 to 60 latitude band.

Davis: I'm thinking of the projections from the Manabe-Wetherald weather model only. But let me return to what I said earlier. Just wait long enough and you will get as much temperature rise as you want.

Clark: All right. But waiting long enough to have very large temperature increases isn't a useful contribution to a management and policy orientation discussion. How much by when is important. We should make a mental footnote that when we are discussing this worst case upper bound, we should indicate whether it will be in the middle of the next century or some time in the indefinite future. Five degrees isn't the relevant number for most of the reference points that have been implicit in the discussion around the table today. It may be a thoroughly defensible number for certain specified times in the distant future.

Davis: The problem is that whenever you make a precise estimate, then you start this argument about the accuracy of the models. We spent about an hour yesterday morning discussing the validity of the models which isn't something we can actually deal with in this group.

Clark: If your number had been a range of one to five degrees I would have no quarrel. It's locking on a given number and sanctioning it by repetition that is dangerous.

Davis: You suggested two degrees. I don't think that's helpful either. The range from one to five is okay.

Jones: A range of one to five can make a huge difference in terms of impact on insect populations.

Clark: That's why the strategy that scientists are now struggling to put into place should be one of dealing with uncertainty.

Davis: From a policy point of view, it's useful to know that there is that degree of uncertainty. What the argument is about is how expensive the warming is going to be relative to how much it will cost to do anything about it. It's useful to know that a two-degree warming would cause some changes but nothing drastic. A warming more than that, as far as insects are concerned, can be serious. That's a very important distinction to make.

Clark: Absolutely. If somebody could estimate the form of the relation between pest virulence and temperature, and decide that the curve has a large kink in it, that's the information the policy debate is madly looking for. I don't think that science is going to support that sort of thing. But certainly, any subfield that can come up with such a relation will be doing an incredible service in focusing the debate on a relevant issue.

Ruttan: An important generalization is coming out of the discussion in this section. Beginning about 25 years ago, certain materials became so cheap -- fertilizer, for example -- that we used the materials as substitutes for knowledge. People were expensive, so we let the materials substitute for them. This took place in both crop and animal nutrition. What we're saying now is that, whether you think of it as knowledge intensive, information intensive, or management intensive, we are in a situation where materials again are becoming expensive, either to the individual farmer or to society. It tells us that we should substitute knowledge for resources in some sense. This is the general principle.

Rawlins: I agree.

Ruttan: The second question is, how do we get that knowledge on the ground where it can be used? It seems to me that in a society where people are very expensive, it is economical to use machines to get the knowledge used. In a society where people are less expensive, or even cheap, you're going to have to make a much stronger effort to get that knowledge into the hands of people. If everybody is going to do it on the back of an envelope, they are going to have to understand the principles. Extension programs in developing countries will have to go beyond teaching practices to teaching principles. Unless you can teach the principles, then people are not going to be able to do the calculations.

Rawlins: And that might be more costly.

Ruttan: That kind of extension is very manpower intensive. It's the kind of thing that the World Bank is trying to do with its Training and Visit system. But extension workers in developing countries are very cheap.

Waggoner: The approach you're talking about is such a good idea-- why isn't it in use? What is the experience? Is it getting done or do we just keep saying we ought to do it?

Ruttan: It's very controversial. The World Bank T & V system is very manpower intensive. Many people are worried about whether the costs will be sustainable when external support is no longer there. We are talking about a world in which we're going to have to be information intensive or farmers just aren't going to be able to get the productivity out of the technology that should be available to them. The level of rural schooling in a wide range of countries -- from Brazil to Pakistan -- is inexcusably poor. Until they do something about under investment in the schooling of rural people, the productivity inherent in the technology will be unrealized.

Cheng: We don't need to accuse others; our own approach to teaching farmers is often too simplistic. For instance, just last week I heard a very disturbing thing in a neighboring state about nitrogen recommendations. Instead of teaching principles, they are coming out with a very simplistic guideline, such as, if you have less than, say, 2 percent organic matter, you can apply fertilizer up to 200 pounds per acre. If it's 2 to 4 percent, apply 160 pounds. This advice totally ignores the environmental impact.

Waggoner: Why are they doing such a thing?

Cheng: I'm asking our extension people if we are doing the same thing. Are we falling into the same trap? We need to raise these questions about our own programs.

Rawlins: I agree that our society will have to be more knowledgeable and information intensive, but all the knowledge and all the information does not need to be in the farmer's head.

Ruttan: Some of it he can buy.

PART FOUR

Responses to Climate Change

16

Nature Myths and Policy Design

Steve Rayner

Ruttan: I asked Rayner to think about the implications of the kinds of uncertainty we face in the environmental and resource area for agricultural research.

Rayner: In a way I feel that I'm a bit of an anomaly at this meeting. I don't have the background in agriculture or some of the appropriate allied disciplines that most of the participants represent. In fact, my initial training, before anthropology, was in philosophy and theology. It may be appropriate, however, since philosophy and theology are concerned with the problems of faith and reason. I have faith that we could see a 4-degree temperature rise sometime in the next 100 years. But reason tells me that the climate models aren't sufficiently good at the moment to let us know if the 4 degrees are Centigrade or Fahrenheit.

This kind of variation reminds me of the question that was asked yesterday about the optimal climate. It reminded me of a story I heard back when I was a philosophy student about a young man who was charged to go home over the weekend and to write an essay on the topic, "How Do I Know I Exist." He returned on Monday looking very haggard, unshaven, and worried, and said to his professor, "Professor, you've got to help me. I haven't slept all weekend. I really worried about it. Do I exist?" And the professor said, "Well, who wants to know?"

That's the question we haven't asked yet. Is climate optimal for what? And resources for whom? So I want to introduce the question of who wants to know at this stage because, in a sense, what I'm proposing is that, having started with a topic entitled "Resource and Environmental Constraints," we're adopting a nature-centered view of the problem. We could equally have developed an anthropocentric view of the problem. We could have started out asking what the institutions are, what are their resources, and how are they threatened by change. Resources are recognized or not recognized according to a variety of societal and institutional variables. One man's hazardous or noxious waste could be another person's valued resource, depending on whether it's piling up in a stable or spread out over the fields. Clearly some locational implications are involved. There's also a question of activities and interests. Where I dwell at the moment in east Tennessee, strip mining is an important economic activity. What you have been talking about in the past two days as "top soil" is called "overburden" by the mining industry. The terminology puts a different complexion on the nature of the resource.

One thing that emerged over the last two days as the most important resource constraint is human resources of institutional information and capacity, which are defined very differently. Climate change, in fact, can be viewed as a resource rather than as something that only impacts on resources. Certainly, if you're in the tourist industry, climate is a very important resource.

Anthropologists like to tell stories. I want to tell you about the Lele and the Bushong, two tribes who live in what is now Zaire. They live on either side of the Kasai River. The Bushong say that their summers are terribly hot and difficult. The Lele describe their summers as very leisurely and pleasant. The standard climatological data gathered by the ethnographers of these two tribes, doing their research simultaneously in the 1950s, indicated that such measures as temperature and rainfall were identical on both sides of the river. There was no difference. So what caused the tribes to perceive climate differently? The most important resource in both societies (forgive me the sexist construction here) is women. In anthropology, kinship is commonly discussed in terms of the exchange or competition for women. The other two relevant resources were raffia cloth and cattle. The Lele pay their bride wealth in raffia cloth, which is gathered by women and woven by men. The Bushong pay bride wealth in cattle. To get access to the resource of women, you must possess either raffia cloth or cattle. Now, the other twist is that the Lele are polygamous: one man, many wives. The raffia is woven by the old men, so the old men have a monopoly on the resource of women. There's not much young men can do to get a wife except to wait until they're older. So, in the summer, they sit around under the palm trees drinking palm wine and enjoying the pleasant climate. The Bushong, on the other hand are monogamous. They pay their bride wealth in cattle. A young man who wants to get a woman works very hard through the summer trying to increase his herd size so he can afford a wife.

You can see how the incentive structure and the functional activities cause the two societies to have different views about what is an optimal climate. I've used an exotic example but there are others that are more familiar. Energy efficiency was not viewed as a resource by utilities until very recently because they emphasized demand expansion. Now that attention has shifted to the supply side, utilities are actually paying energy service companies for the avoided costs of building new plants to generate more electricity. Efficiency is now seen as a resource. We've had a transition in the perception of what is recognized as a resource.

At least three distinctive policy responses can be made to climatic change. Each has a moral component. Each is supported by a distinctive myth about the nature of resources. The first policy response is the *preventivist*. Just say no to climate change. This view is supported by a naturalistic philosophical perspective that holds it is morally wrong to mess with nature. The myth holds that nature is fragile and natural resources are scarce. Any slight perturbation will irretrievably upset the balance of nature.

There are some fascinating parallels here with the religious right on the issue of sex education and with the political left on the issue of civil defense. The religious right, at least in my part of the Bible belt, insists that we shouldn't teach kids about sex because they'll be encouraged to experiment with it. Similarly, we find that certain groups on the political left argue that we shouldn't even talk about civil defense because this will encourage people to accept the dangerous myth that nuclear war is survivable. We are told by hard line preventivists that we shouldn't be talking about adaptation because this will encourage people in the notion that we can adapt to climate change. We should be concentrating on prevention.

The second approach is *adaptivist*. This approach is illustrated by the people who erected the statue to the boll weevil. It may be no coincidence the town is known as Enterprise. They saw change as an opportunity. And just as the boll weevil presented an opportunity for the people of Enterprise to change their farming system, climate change may well present us with opportunities to recognize new resources and to use them in different ways. Related to this position is the moral judgment that it's wrong to curtail development. For example, the position that a preventivist strategy will condemn the poor peoples and

158

poor countries of the world to a state of continued poverty or underdevelopment is seen as morally wrong. The underlying nature myth perceives nature as robust. It does not become unbalanced easily.

These two myths go very far back in American society. It was represented in the arguments over the management of national forests at the turn of the century. More recently, though, we've had the *sustainable development* response. This is a sort of Hegelian synthesis of the classic dialectic. The notion is that growth should be controlled. The moral imperative is to preserve choice for future generations. There is an image that nature is robust within limits. A certain amount of perturbation can be tolerated, but one must be careful not to exceed the limits -- even though we don't know what the limits are. These myths are of particular importance when we're dealing with issues that are attended by extraordinarily high uncertainty, and where the stakes are both high and long term. Three types of uncertainty have been identified by people looking at these issues. *Technical* uncertainty is what we are talking about when we look at the uncertainty bands around an estimate or a measurement. That very image gives the notion that uncertainty is always reducible. As one moves from bench science into the broader environmental realm, we frequently encounter *methodological* uncertainty. Do we even have the appropriate tools for modeling and dealing with a problem? Finally, there is *epistemological* uncertainty. Epistemology is the study of whether you have appropriate conceptions for dealing with a problem. Clearly that's an issue that affects us in the kinds of debates we've been having about the extent and nature of climate change.

Three kinds of stakes are involved: local, societal, and global. Different tools are appropriate for dealing with them. When you have low stakes and merely technical uncertainty, you pretty much know what the probabilities are. However, when you move to methodological uncertainty and medium decision stakes, you're moving into much more uncertainty. The probabilities are less tractable. Here, decisions must be more in the clinical mode. However, when we then get up into the area of high stakes and epistemological uncertainty, we're in a realm of indeterminacy. Very often, as in the climate area, we don't even know what the signs of changes will be. For example, there is still considerable debate whether cloud cover will provide a positive or negative feedback for the climate system.

If you can't increase precision, what do you do? You try to have a prudent response, analogous in some respects to insurance. The catch here is that the advocates of the different nature myths which I talked about earlier have a different approach to the issue of prudence. The preventivists focus on avoiding the worst case costs. They are willing to spend a lot of money on insurance that may not be needed. This can create considerable societal discord.

I want to try to distinguish between two types of resource impact that we might want to consider in further research. We have a project at Oak Ridge in which we're reviewing the programs that U.S./AID has in place to determine whether they ameliorate or exacerbate the issue of climate change. What we tried to do in a first cut was to distinguish two kinds of resource impacts: the long-term secular changes and the sort of short-term emergencies that are likely to come along before we get to whatever temperature changes may be occurring in the middle of the next century. The interesting thing is that we're already seeing the cumulative effects of urbanization, transmigration, and others. Our view is that it makes perfectly good sense to discuss them as if they were problems of decision under uncertainty and possibly under risk. We already have the tools to deal with these problems.

The longer term issues are likely to fall into two areas. One is industrial metabolism, particularly energy use, and the other will be land-use change on a global scale.

We spoke yesterday about the importance of energy conservation in the United States. Biomass is considered -- by Oak Ridge researchers at least -- a serious contributor to the very long-term reduction of greenhouse gas emission. By the year 2,000, the developing countries and newly industrialized countries will contribute more to CO_2 emissions than the industrial world. We feel that biomass technology is going to be very, very important for these developing countries if they are going to avoid more intensive use of coal -- the energy equivalent of heroin. And it may also be worth our while to consider whether the United States would get a bigger bang for its buck in terms of CO_2 reduction by developing and transferring biomass technologies to developing countries. We can try to reduce our own emissions through accelerating replacement of the existing plant-generating stock in the United States. But we actually need to think about the interaction of agricultural and energy systems on a global scale. Unless we get a serious handle on global emissions problems, it won't make much difference whether we close down our coal-fired power stations in 10 or the next 40 years.

Implementation of a biomass program raises the whole question of market failure. We've had a lot of research on market failure in energy efficiency research. Crosson raised the issue yesterday when he asked why, if energy efficiency technologies are available, they aren't being

160

introduced. We've identified several reasons. One is the high, first cost problem. Highly efficient compact fluorescent bulbs are available to replace incandescent lights and to save considerable amounts of money and energy over the life of the bulb. They cost $25. When you stop at the hardware store on Saturday morning to replace a light bulb, you may not feel like splashing out $25.

The other problem is split incentives. The people who, in particular, occupy commercial and industrial buildings are generally not the owners, and certainly were not the builders. Owners and builders have an incentive to install the lowest first-cost appliances in buildings, not the ones that are most efficient over the life of the building.

Let me now turn to the issue of land-use transformation. At Oak Ridge we are trying to bring together people from the social sciences and the biophysical sciences to work out integrated models of how resources are perceived, used, and transformed. At the moment, we're simply trying to develop simple algebraic models of the various parts of the system. We're not at the point yet where we've developed any computer simulations. The time has come to move away from seeing land-use change in traditional social science or biophysical terms.

In closing I want to address the broader issue of information flows. Yesterday, Ruttan suggested that I talk about international policy. The traditional model of international relations was relations among governments. Nation states would make their decisions according to their own national political cultures and agendas. They would then come together at the governmental level, formulate some kind of consensus, and embody it in a treaty, which they would then be individually responsible for enforcing within whatever framework of international law applied. This was very much the model that was followed in the attempt to ban CFCs in aerosols. The aerosol ban, one of the precursors to the Montreal protocol, failed. It was also the model for the negotiations around the U.N. convention on the Law of the Sea, to which the United States eventually did not become a signatory. In this model, it's assumed that the decision makers know what the national self-interest is when they go into the negotiations. The assumption is that the science is already pretty clear.

There was a dispute among the E.C. countries, United States, Canada, and the Scandinavian countries, over the science of CFCs and aerosols. When this dispute is described in the United States it is usually in rather cynical terms to the effect that the Europeans were simply interested in stepping into the market niche that the Americans might vacate through decreasing CFC production. However, if one

looks at how science is incorporated into governmental decision making
in different countries, one can see that, in fact, a lot of the disagreement
was associated with the problems of how the different societies dealt
with uncertainty in science. The United States and the Scandinavian
countries were moving toward avoiding the worst-case scenarios whereas
the European Economic Community countries were concentrating much
more on avoiding the opportunity costs. Given the failure of the aerosol
ban, there was an explosion of direct networking among non-governmental organizations, such as scientific, technical, and industrial groups
across national boundaries. We had things like the development of the
Council for Responsible CFC Use. American environmentalists went to
talk to West Germans, and West Germans talked to the United
Kingdom. There was a proliferation of information flows directly across
national boundaries. By the time of the Montreal meeting, a consensus
on a course of action had been agreed on. It was necessary only to go
through the process of symbolic confirmation: which is what the
Montreal protocol amounts to. It's not the means by which we're going
to reduce CFC emissions. As an anthropologist, I think symbols are
very important, of course. But if we were serious that the Montreal
protocol would have the effect of reducing CFC emissions, we could not
have allowed the Soviet Union to become signatories and to open new
CFC factories, which, in fact, we agreed to do. But the people who
were involved were smart enough to recognize that, in the long run, it
is much more important to have the Soviets involved, even symbolically,
than to quibble over details. I have described this use of specific kinds
of expertise to deal with global problems as "thinking globally and acting
locally." My colleague, Luther Gerlach, who is in the Anthropology
Department at Minnesota, has pointed out that the ability to think
globally is itself a very specific kind of local knowledge that is restricted
to quite a small community in the world.

Let me now turn to climate change. The United States opposed
participation in the Intergovernmental Panel on Climate Change (IPCC)
until the Bellagio meetings, at which the SCOPE project, one of those
international scientific organizations, turned its attention to policy issues.
The United States and other governments sensed that the lead was
being taken by the World Meteorological Organization rather than by
governments. The Bellagio reports were policy documents without
government fingerprints. Governments felt that they had to deal
themselves back in. An Intergovernmental Panel on Climate Change
was formed which involved much of the expertise that previously was
involved in direct interactions across national boundaries. That

expertise is now being redirected back into the more traditional model of intergovernmental decision making. We have a hybrid between the traditional model that I first described and the polycentric model.

The reason I have discussed this example is because in the agricultural area, we have a whole series of cross-national institutions -- the Consultative Group on International Agricultural Research (CGIAR) and others -- which are capable of collaboration across national boundaries without having to go through government processes. These organizations may be one way to intervene in countries where the governments very often are themselves a very large part of the problem. Part of the solution is the development of non-governmental organizations in developing countries which can interact directly with our non-governmental organizations to transfer knowledge and technology and to conceptualize issues in a manner that induces changes in policy.

Waggoner: What's the value of attempting to narrow probabilities with respect to future climates?

Rayner: I didn't talk about Coase (1960) and Rawls (1971). Their problem is, "How do you seek a fair solution to a problem, such as the intergenerational problem, in which there are both winners and losers?" Coase, an economist, argued that the way to get a fair solution is to reduce uncertainties to an absolute minimum so you know exactly who's going to win by how much, and who's going to lose by how much. You then allow the participants to negotiate a redistribution in which either the losers bribe the winners not to win, or the winners compensate the losers for the loss. Rawls, on the other hand, argued that that won't work for a variety of reasons, including the fact that you never have equal market power on both sides. Rawls suggested that one way to get a fair solution, particularly when you can't know who the winners or losers are going to be, is to ask what solution each player would chose in the absence of knowledge about which side he or she will be on. That gets you a fair outcome.

Waggoner: Does that mean shut down all sources of CFC?

Rayner: It suggests that we do not pursue greater precision in scientific knowledge until we figure out how it relates to the equity issues. On the other hand, Coase is saying that the scientific knowledge must be very precise in order to have an equitable solution. The problem is that neither course of action is achievable.

Waggoner: But we are always faced with a choice of where we're going to put our chips in allocating our science resources.

Rayner: I would pursue the Coasean solution to short-term emergency issues for which we have enough reasonably good information. For the long-term problems in determining issues, put the money into increasing institutional and societal resiliency so that we can better respond to great certainty as more knowledge becomes available. I suggest pursuing the Rawlsian strategy for the latter category.

Chen: The CFC case seems to be one in which there is enough information for most climatologists to agree that change is coming. This provides the motivation to want to negotiate. But there is no agreement on the regional distribution. There is, therefore, uncertainty about winners and losers. Do you think that the disagreements are about the science or because of biases about winners and losers? You don't have any cynical view of why some people are claiming uncertainties, do you?

Rayner: I don't think that people are deliberately and cynically manipulating uncertainties, giving their own views an advantage. But I do think that we have these underlying myths about the nature of nature that introduce ideological considerations into policy preferences.

17

Research Resource Allocation

Stephen L. Rawlins and C. Eugene Allen

Rawlins: I've been stimulated by Rayner's discussion, particularly the point that before we decide what to do, we ought to think about who needs the answers. One problem agriculture faces is that it does not have a single customer. Sonka discussed food security. Is that the central issue agriculture should address? If so, it is the consumers of food that we should be concerned with. Or is it the producer and the profitability of his enterprise? Or should we be concerned with the environment and the health of rural communities?

Ruttan: The decisions are those that would be more relevant to research decision making at the CGIAR, USDA, state experiment stations, and granting agencies. I guess one way I would ask the question is, given what we think we know about the next 20-40 years, and having some sense about the uncertainties, how should we change our research portfolio? Part of that portfolio might include research to improve our knowledge about the future. It might also include greater effort to monitor changes so that we know what is happening and in what direction the changes are taking us.

Rawlins: Maybe my cynical response would be that you are suggesting that we should be primarily concerned with ourselves.

Ruttan: No, we're concerned with the producers and consumers because they put the money in our pay check.

Waggoner: The word portfolio is important.

166

Rawlins: Certainly, that is true. I suppose we could divide our multiple concerns into two broad categories: (a) the impacts of the environment on agriculture; and (b) the impact of agriculture on the environment. By environment, I mean everything external to the agricultural system, including political and social as well as physical, chemical, and biological elements. The framework developed by the IPCC provides a useful division of research responsibilities: science understanding, impacts assessment, and response strategies. One thing is clear. Agricultural research needs to interact with the rest of the scientific community.

Looking specifically at research related to global environmental change, agriculture has contributions to make in all three IPCC categories. To develop a more complete understanding of the system, agriculture can make better measurements of biogeochemicial fluxes of greenhouse gases from agriculture, contribute to a better understanding of the biosphere component of the carbon cycle, and bring our understanding of soil and hydrology to bear on the inputs needed to improve GCMs. In the area of impacts assessment, I was intrigued with Sonka's suggestion for linking the biophysical land use models with models of the socio-economic system. Population is an important driver in this system that cannot be ignored. If we deal only with physics, chemistry, and biology we'll fall short of coming up with answers that are needed. Only after we understand the system and its interactions, and the social and economic impacts of environmental changes on the system can we develop rational response strategies and policies. Unfortunately, much of what we are doing now is not directed toward the ultimate development of policy. We need to create these linkages. Making an attempt to outline policies now will help to reveal the specific knowledge gaps that need to be filled. Response strategies should include both strategies to help prevent negative environmental changes and strategies to adapt to changes by increasing the resiliency of agricultural production systems.

Finally, we must be aware of the critical relations between water and agricultural production. Water is the lifeblood, and the most frequently limiting factor in agricultural production. Not only is agriculture the largest user of water -- if you include forests and rangeland -- it also is one of the major sources of water. How these resources are managed can have a substantial impact on the nation's water supply.

Ruttan: Water has been a central issue in these conversations.

Rawlins: It is rapidly becoming the most limiting environmental variable as far as agricultural production is concerned.

A third research priority should involve the design of response strategies. They should include both prevention and adaptation. We ought to be very seriously concerned with developing greater resiliency within our food system. Agricultural research also needs to be involved in the design of strategies to help prevent global warming. A first step is to more carefully assess the impact of agricultural systems on the radioactive gases that are emitted from agriculture. After we assess and measure it, we need to develop practices that will minimize emissions, particularly methane and carbon dioxide. Agriculture has an opportunity to be on the positive side of carbon flux by sequestering it. There are opportunities through reduced tillage to increase organic matter in the soil. We must be involved because agriculture occupies such a large part of the land area.

Allen: This conference is a good example of driving home concerns with agriculture and the environment that must come closer together on both the input and output sides.

The concept of sustainable agriculture will become increasingly important in the decades ahead. But it is not as widely accepted at this point as it should be. I've used the example of holistic medicine as an analogy. Holistic medicine is a very appropriate concept in thinking about health. But it is not an acceptable term in the health professions because, in the past, it was associated with quackery. The concept of sustainable agriculture is not yet that contaminated. It was promoted originally as only organic or chemically free, but this is only one aspect of it. Other terms such as "alternative" or "low-input" agriculture have been used, but "sustainability" captures more of the concept at the intuitive level. An agriculture that is sustainable must also be profitable. If we can accept the concept of sustainability, it can be used to provide a philosophical foundation for our research priorities.

The other thing that is important for our research programs to recognize is that there's not just one kind of land use in Minnesota, the United States, or the world. In Minnesota, for example, we devote approximately 20 million acres to cropland, 18 million acres to forestry, and 5 million acres to recreational uses. There are 7 million acres of peatlands. About 5 million acres are in multiple use: forestry and recreation. Agricultural land uses have undergone significant change. Land used for crop production has declined in the northeast. But maize and soybeans have moved north in the west and northwest parts of the state. The structure of agriculture is increasingly bimodal. We have both small farmers and large commercial farmers whose needs must be addressed. The distribution is more bimodal than it was 20 or 25 years ago.

When one also considers the need to bring together the agriculture, environmental, food safety, water quality, profitability, and trade dimensions, the system becomes exceedingly complex. We need good disciplinary sciences. But we also must go beyond the reductionist mode and learn how to integrate our knowledge and our technology. One thing I am very excited about in the initiative for agriculture research by the NAS/NRC Board of Agriculture, is that 40 percent of the new funding is proposed to go to interdisciplinary teams. We believe this is not only needed for applied agricultural research but, also, in many areas that have been funded primarily by single-investigator disciplinary grants.

Another point I want to make is my concern that politicians are making decisions about science without the needed scientific input. A problem we have in this country, at least in my view, is that we're thinking and acting more locally. We must bring to the general populace a greater understanding of some of the scenarios that we're discussing if we expect the politicians to change. Just as an example, too few of our undergraduates today are coming out of universities with a general sense of the issues that relate to food and hunger, natural resources, or the environment.

Participants' Perspectives

Ruttan: Let's now see what the others would like to put on the agenda.

Waggoner: We must learn how to choose amongst the possible abatements and adaptations Sonka spoke about so clearly. Then we must find the obstacles that are stopping us from doing the things we know how to do. We must continuously develop and test and adapt crops and systems outdoors. We must quantify the effect of land use on the parameters of carbon and water exchange. This is not only important as an input to global circulation models but it will also determine the limits that can be put on agriculture by those attempting to slow the climate change.

Rosenberg: In the article Crosson and I wrote for *Scientific American* (1989), we tried to get a handle on the kinds of environmental deterioration that are caused by agriculture: desertification, salinization, and erosion, for example. We were both extremely frustrated by the poor quality of data on which we could draw. The knowledge base on land use and land degradation is woefully inadequate. We have to find how to improve the way we characterize and document the magnitude of environmental problems.

Coming at it from another side is the International Geosphere-Biosphere Program. I served on the first IGBP committee. One idea that I thought was most useful at the time came to be called the "Geosphere Biosphere Observatory." It proposed a network of stations where important observations of land change process could be made systematically and over long periods of time. Scientists in the developing countries could be enlisted to work on monitoring of natural resource problems, such as erosion, salinity, and desertification. These are immediate problems that should be able to maintain the interest of LDC governments.

Chen: I want to follow up on the interdisciplinary research issue. There are three areas of concern: One is the problem of working at the natural and social science boundary. The second is the interface between climatology and agriculture. I organized a meeting in February at IIASA on the issue of using climate scenarios in impact studies. A range of issues here need a lot more technical work. The third area is the food security-hunger nexus. A systems view will be very important for determining the constraints on sustainable agriculture.

Rayner: I'm still not quite sure why in an institution like Oak Ridge, which is predominately engineering and natural sciences, the social sciences also seem to thrive. One thing I would like to emphasize about interdisciplinary research is that it is very difficult. We have found, for example, in our land-use project that you can talk with colleagues from other disciplines for several hours or days thinking you understand each other then, suddenly, discover that you've been talking past each other. A lot of patience is required. It will be a challenge to the way we do science, both in our laboratories and our universities.

As far as a research agenda is concerned, the issue of biomass for energy would be high up on my agenda. And, as I've learned at this meeting from Rawlins, the issue of biomass for lignocellulose food sources could be very important. It may be particularly important for developing countries and even in this country as a source of cattle feed. Something like a hundred million tons a year of grain in this country is used to feed livestock.

In the institutional area, though, which is the one in which I feel that I have the most competence, I would like to see a focus on resource management. Attention to the issue of market failure and institutional design are also important. We know, for example, how price support systems distort markets. We have seen in recent years a very strong emphasis on the private sector in developing countries. Sometimes this has been productive, sometimes it has been counterproductive. There has been failure to recognize that common property systems that are capable of effectively managing resources do indeed exist in different parts of the world. We are, in fact, in danger of perpetrating a loss of institutional diversity. In other words, we're losing a lot of the small-scale institutions that we could learn from to understand how to handle the big-scale problems better. The second of the three models of international decision I referred to earlier was actually derived by Luther Gerlach and myself from decision making in African tribal societies.

Finally, it makes perfectly good sense to develop policies that are capable of responding to short-term local impacts of climate change. I am referring to issues like flooding, refugees, epidemics, and others.

Jones: We should concentrate efforts on pesticide alternatives. This is a simplistic approach to the water-quality issue. If you could just eliminate the pesticides, then outside of nitrogen you've solved water-quality problems. Of course, we can't do that overnight. But I think we have to be careful with the water- quality problem not to make another acid rain case where you know what needs to be done but don't do it.

We have to emphasize integrated management approaches, not only integrated pest management, but crop management programs. There's a lot going on in biology in the high tech area, but there's been a decline in the nuts and bolts stuff. Since 1971, the number of entomologists employed by the state and federal level has decreased by 30 percent. And the same thing is happening in the other biological fields.

Another area I want to comment on is the assumption that Third World countries should implement environmental programs using the same criteria we use in the United States. An entomologist, B.D. Walsh, wrote in 1866, "Let a man profess to have discovered some new patent medicine and people will listen to him with attention and respect. But tell them of any simple common sense plan based upon scientific principles and they will laugh you to scorn." Not a lot has changed since 1866. We are imposing our risk-benefit parameters on Third World countries where it is likely to be very difficult.

Bochniarz: We now need social science research to be more concerned with understanding institutional diversity, particularly political institutions. On the question of regulation and deregulation, we need to introduce an international perspective that draws on more diverse experience. The problem of internalizing externalities is perhaps largely unimportant.

Munson: We need to figure out better ways to increase efficiencies of input use. Water is without a doubt the most limiting factor in our production system. We need to take a much closer look at the water use efficiency of various crops and determine what we can to do to improve it.

Larson: I agree with a lot that's already been said. We need to inventory our natural resources so that they can be identified on a spatially accurate basis, and then to develop data bases that characterize these natural resource units.   We need to bring these data bases together for use in development, management, and impact assessment. This contrasts with the trial-and-error method of research often used in agriculture.

Cheng: We need to achieve closer articulation of the sciences with the social sciences and the humanities. I recently asked someone in the humanities if he ever thought of agriculture as part of a culture. Agriculture is different in different countries because the culture is different. And agriculture in turn affects the culture. His first reaction was quite negative but after we talked a bit, he became interested enough to want to incorporate the idea into a world-culture course.
Rayner's comment about the decreasing institutional diversity reminds me of the experience my friend, Jim Cook, had in China. He went there in 1978 with the first plant pathology delegation. When he came back he said, "You know, the Chinese are practicing integrated pest management!" Their technology, perhaps suited to their particular agricultural system, may not be suited to our system. During the 10 years I've been going to China, I am worried that increased use of pesticides has almost totally eliminated the traditional concept of integrated pest management. Another area of technical knowledge that

was almost lost was crop rotation in the United States. With the introduction of chemical inputs, we went to continuous corn and continuous wheat. When we started looking back, we found that we had been misled because the yield increases have masked the deterioration in soil quality.

Davis: Two new sub-disciplines are developing within the field of ecology that are relevant to a lot of the discussions we've had here. One is landscape ecology. The other is conservation biology. The emergence of these fields has resulted in the formation of new scientific societies and new journals. There are new courses and new graduate curricula. Certainly, ecologists are concerned with natural landscapes but they recognize that landscape is very much shaped by human activity. Many problems discussed here fall within the purview of landscape ecology. Similarly, human impacts are creating many problems in conservation biology; some interesting basic scientific issues must be addressed in developing strategies for conservation of species.

The need for basic ecological research at the interface of human and natural impacts on the landscape has resulted in the formation of a new journal called *Ecological Applications*. It is trying to deal with the literature that falls in the area between applied biology and basic biology. Much of what we have discussed is actually systems or ecosystems research. These new developing fields of landscape ecology and conservation biology are evidence that ecologists are moving in your direction. But are you moving in our direction? For instance, I teach a course in ecosystem ecology. Faculty members from Agriculture come in as guests to present about 10 percent of the lectures in the course. I've had students from forestry, wildlife, and fisheries but I don't believe I have ever had a student from any agriculture discipline. Yet training in ecological research is essential for students in agriculture. Most students in ecology are very much interested in problems such as those we've been discussing here. Some of my students, for instance, have taken a course in tropical agriculture. I think we can easily establish better communication than we have now.

Abrahamson: I agree with Rawlins that agriculture has not been very sensitive to environmental concerns. I was pleased to hear some sensitivity is developing, at least if this group is representative. On the other hand, I've been very uneasy during our conversation. I grew up on a Minnesota farm. I remember when the first tractors were bought and I helped to bring home the first bag of fertilizer. Then I spent nearly 20 years working as an applied physicist. I enjoy machines and technology, but when I sit here and hear this kind of high tech agriculture talk, I get nervous. I just don't like management. And what we're talking about here is turning the whole world into a zoo. Conservation biology is coming along just in time to study natural systems as they go down the tube.

Clearly I have a great deal of the preventivist theology that Rayner described. But the choice is not between adapting and limiting. If we're going to avoid what I think will be catastrophic changes down the line, we have to limit emissions of the gases and we have to do it vigorously. That means limiting fossil fuels consumption and ending deforestation. Fossil fuels account for about 60 percent of the greenhouse gases. We heard today that deforestation accounts for someplace between 10 and 25 percent. But, also, we must adapt to or cope with those very large changes that are unavoidable. Now, the costs of coping are very high. Even though you can go a long way with technical fixes, it will not be enough. It's necessary but it's not sufficient. The required changes will demand true grass roots political support and public understanding to marshall these resources. That's why I'm an advocate. I do advocacy because I just don't see any way to marshall the resources and political support to deal with these issues unless there is a public that understands the implications of inaction and of not providing necessary resources.

Sanchez: We should by all means get away from the extremes of being too catastrophic or too utopian. We need more emphasis on the abatement technologies. But abatement technologies have been around for some time. The question, that Paul Waggoner and Pierre Crosson kept raising -- how come technologies are not utilized -- has not been answered. We should focus more on the technologies that have a win-win potential, that both increase production and have positive environmental gains.

My second point is that I'm just fascinated by the efforts to put some realistic economic values on environmental costs. The problem of internalizing externalities calls for some institutional innovations.

There were several very important observations made in the last several days. H.H. Cheng observed that methane emission from paddy rice is likely to occur only at very low redox potentials. Most of the rice fields will not have those low redox potentials. What is going on? The people who work in methane certainly should look at the chemistry more carefully.   Larson's comment that desertification is reversible is very important.

Agronomists and other agricultural scientists are turning to ecology in a serious way.   At my own institution, our soil scientists, entomologists, and foresters are taking courses in ecosystems.   Conservation biologists are learning about soils, plant and animal science.   They are also learning about the future of agriculture.

PART FIVE

Sustainable Agriculture and the Future

18

Issues and Priorities for the Twenty-first Century

Vernon W. Ruttan

When we look even further into the next century, there is a growing concern, as noted earlier, with the impact of a series of resource and environmental constraints that may seriously impinge on our capacity to sustain growth in agricultural production. One set of concerns centers on the environmental impacts of agricultural intensification. These include groundwater contamination from plant nutrients and pesticides, soil erosion and salinization, the growing resistance of insect pests and pathogens and weeds to present methods of control, and the contribution of agricultural production and land-use changes to global climate change. The second set of concerns stems from the effects of industrial intensification on global climate change. It will be useful before presenting some of the findings of these conversations to briefly characterize our state of knowledge about global climate change.

There no longer can be any question that the accumulation of carbon dioxide (CO_2) and other greenhouse gasses -- principally methane (CH_4), nitrous oxide (N_2O), and chlorofluorocarbons (CFCs) -- has set in motion a process that will result in some rise in global average surface temperatures over the next 30-60 years. Substantial disagreement is evident about whether warming due to greenhouse gasses has already been detected. And great uncertainty continues about the increases in temperature that can be expected to occur at any particular date or location in the future.

The several greenhouse gases differ with respect to (a) their radiative properties, (b) their different lifetimes in the atmosphere, and (c) the extent to which they undergo chemical transformation into other substances. Estimates reported by the U.S. Department of Agriculture (see Fig. 2), based on radiative properties, suggest that carbon dioxide accounts for roughly half the radiative forcing of global climate change. Other estimates, which take into account the different lifetimes and the chemical transformations, project a larger contribution from carbon dioxide and a smaller contribution to radiative forcing of climate change than the approach employed by Hanson (Lashoff and Ahuja, 1990; Nordhaus, 1990).

Most carbon dioxide emissions come from fossil fuel consumption. Biomass burning, cultivated soils, natural soils, and fertilizers account for close to half the nitrous oxide emissions. Most known sources of methane are a product of agricultural activities, principally, enteric fermentation in ruminant animals, release of methane from rice fields and other wetlands, and biomass burning. Estimates of nitrous oxide and methane sources, however, have a very fragile empirical base.

On a regional basis the United States contributes about 20 percent and western and eastern Europe and the USSR about 30 percent of greenhouse gas radiative forcing. In the near future contributions to radiative forcing from Third World countries are expected to exceed that of the OECD and the former centrally planned economies. Calculations based on radiative properties of several greenhouse gasses suggest that land use transformation and agricultural production could account for as much as 25 percent of the forcing of global climate change (Figure 2). It is apparent that calculations taking into account the different lifetimes and chemical transformation of the several greenhouse gasses would attribute to a somewhat smaller share of climate change forcing to agricultural sources.

During the conversations, Rayner, as well as several other participants, characterized the alternative policy approaches to the threat of global warming as *preventivist* and *adaptionist*. A preventivist approach could involve five policy options: reduction in fossil fuel use or capture of CO_2 emissions at the point of fossil fuel combustion; reduction in the intensity of agricultural production; reduction of biomass burning; expansion of biomass production; and energy conservation.

The simple enumeration of these policy options should be enough to suggest considerable caution about assuming that radiative forcing will be limited to anywhere near present levels. Fossil fuel use will be driven, on the demand side, largely by the rate of economic growth in the Third World and by improvements in energy efficiency in the developed and the former centrally planned economies. On the supply

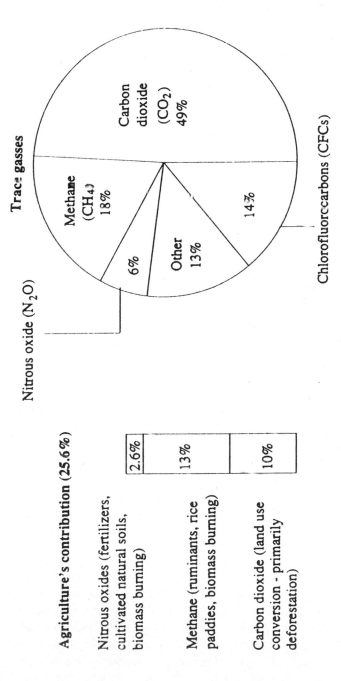

Figure 2. Contributions to increases in radiative forcing in the 1980s. Adapted from J. Reilly and R. Bucklin, climate change and agriculture in U. S. Department of Agriculture, World Agriculture Situation and Outlook Report (Washington, DC: USDA/ERS, WAS-55, June 1989), p. 44.

side it will be constrained by the rate at which alternative energy sourcesare substituted for fossil fuels. Of these, only energy efficiency and conservation are likely to make any significant contribution over the next generation. The speed with which it will occur will be limited by the pace of capital replacement. Significant reversal of agricultural intensification, reduction in biomass burning, or increase in biomass absorption is unlikely to be realized within the next generation. The institutional infrastructure or institutional resources that would be required do not exist and will not be put in place rapidly enough to make a significant difference.

This forces me to adopt an *adaptionist* approach in attempting to assess the implications of global climate change for future agricultural research agendas. It also forces me to agree, as Abrahamson has insisted, that we will not be able to rely solely on a technological fix to the global warming problem. The fixes, whether driven by preventivist or adaptionist strategies, must be both technological and institutional.

An adaptionist strategy implies moving as rapidly as possible to design and put in place the institutions needed to remove the constraints that intensification of agricultural production are currently imposing on sustainable increases in agricultural production. Examples would include (a) the policies and institutions needed to rationalize water use in areas such as the western United States and the Indus Basin; (b) management of the use and development of coastal wetlands and shorelands to limit contemporary losses to property and human life; (c) strategies to deal with groundwater management, including the effect of pollution resulting from agricultural intensification. If we are successful in designing the institutions and implementing the policies needed to confront these and other contemporary problems, we will be in a better position to respond to the more uncertain changes that will emerge as the result of future global climate change.

The following research implications emerged from the conversations:

1. *A serious effort should be initiated to develop alternative land use, farming systems, and food systems scenarios for the 21st century.* A clearer picture of the demands that are likely to be placed on agriculture over the next century, and of the ways in which agricultural systems might be able to meet such demands, has yet to be produced. World population could rise from the present 5 billion level to the 10-20 billion range. The demands that will be placed on agriculture will also depend on the rate of growth of income, particularly in the poor countries where consumers spend a relatively large share of income growth on subsis-

tence: food, clothing, and housing. The resources and technology that will be used to increase agricultural production by a multiple of 3-6 will depend on both the constraints on resource availability that are likely to emerge and the rate of advance in knowledge. Advances in knowledge can permit the substitution of more abundant for increasingly scarce resources and reduce the resource constraints on commodity production. Past studies of potential climate change effects on agriculture have given insufficient attention to adaptive change in non-climate parameters. But the application of advances in biological and chemical technology (which substitute knowledge for land), and advances in mechanical and engineering technology (which substitute knowledge for labor) have, in the past, been driven by increasingly favorable access to energy resources by declining prices of energy. It is not unreasonable to anticipate that there will be strong incentive, by the early decades of the next century, to improve energy efficiency in agricultural production and utilization. Particular attention should be given to alternative and competing uses of land. Land-use transformation, from forest to agriculture, is presently contributing to radiative forcing through release of CO_2 and methane into the atmosphere. Conversion of low-intensity agricultural systems to forest has been proposed as a method of absorbing CO_2. There also will be increasing demands on land use for watershed protection and for biomass energy production.

2. *The capacity to monitor the agricultural sources and impacts of environmental change should be strengthened.* It is a matter of serious concern that only in the last decade and a half has it been possible to estimate the magnitude and productivity effects of soil loss in the United States. Even rudimentary data on effects of soil loss production are almost completely unavailable in most developing countries. The same point holds, with even greater force, for groundwater pollution, salinization, species loss and others. It is time to design the elements of a comprehensive agriculturally related resource monitoring system and to establish priorities for implementation. Data on the effects of environmental change on the health of individuals and communities are even less adequate. The monitoring effort should include a major focus on the effects of environmental change on human populations. Lack of firm knowledge about the contribution of agricultural practices to the methane and nitrous oxide sources of greenhouse forcing was mentioned several times. Much closer collaboration is essential among production-oriented agricultural scientists, ecological-trained biological scientists, and the physical scientists who have been traditionally concerned with

global climate change. This effort should be explicitly linked with the monitoring efforts currently being pursued under the auspices of the International Geosphere-Biosphere Programs (IGBP).

3. *The design of technologies and institutions to achieve more efficient management of surface and groundwater resources will become increasingly important.* During the twenty-first century water resources will become an increasingly serious constraint on agricultural production. Agricultural production is a major source of decline in the quality of both ground and surface water. Limited access to clean and uncontaminated water supply is a major source of disease and poor health in many parts of the developing world and in the former centrally planned economies. Global climate change can be expected to have a major differential impact on water availability, water demand, erosion, salinization, and flooding. The development and introduction of technologies and management systems that enhance water-use efficiency represents a high priority because of both short- and intermediate-run constraints on water availability, and the longer run possibility of seasonal and geographical shifts in water availability. The identification, breeding, and introduction of water efficient crops for dry land and saline environments is potentially an important aspect of achieving greater water-use efficiency.

4. *The modeling of the sources and impacts of climate change must become more sophisticated.* One problem with both physical and economic modeling efforts is that they have tended to be excessively resistant to advances in micro-level knowledge in the failure to take into consideration climate change response possibilities from agricultural research, and in the response behavior of decision-making units, such as governments, agricultural producers, and consumers.

5. *Research on environmentally compatible farming systems should be intensified.* In agriculture, as in the energy field, a number of technical and institutional innovations could have both economic and environmental benefits. Among the technical possibilities is the design of new "third" or "fourth" generation chemical, biorational, and biological pest management technologies. Another is the design of land-use technologies and institutes that will contribute to the reduction of erosion, salinization, and groundwater pollution.

6. *Intermediate efforts should be made to reform agricultural commodity and income support policies.* In both developed and

developing countries, producers' decisions on land management, farming systems, and use of technical inputs (such as fertilizers and pesticides) are influenced by government interventions, such as, price supports and subsidies, programs to promote or limit production, and tax incentives and penalties. It is increasingly important that such interventions be designed to take into account the environmental consequences of decisions by land owners and producers induced by the interventions.

7. *Alternative food systems will have to be developed.* A food-system perspective should become an organizing principle for improvements in the performance of existing systems and for the design of new systems. The agricultural science community should be prepared, by the second quarter of the next century, to contribute to the design of alternative food systems. Many alternatives will include the use of plants other than the grain crops that now account for a major share of world feed and food production. Some alternatives will involve radical changes in food sources. Rogoff and Rawlins (1989) have described one such system that is based on lignocellulose, both for animal production and human consumption.

8. *A major research program on incentive compatible institutional design should be initiated.* Large-scale program of research on the design of institutions capable of implementing incentive- compatible resource management policies and programs should be initiated. By incentive-compatible institutions I mean institutions capable of achieving compatibility between individual, organizational, and social objectives. A major source of the global warming and environmental pollution problem is the direct result of the operations of institutions that induce behavior by individuals, and of public agencies that are not compatible with societal development -- some might say survival -- goals. In the absence of more efficient incentive-compatible institutional design, the transaction costs involved in ad hoc approaches are likely to be enormous.

References

Ausubel, H. and E. Sladovich (Eds.). *Technology and the Environment*, National Academy of Engineering (Washington, D.C.: National Academy Press, 1989).

Barnett, H.J. and C. Morse. *Scarcity and Growth: The Economics of Natural Resource Availability*, Resources for the Future (Baltimore: The Johns Hopkins Press, 1963).

Bryson, R. A. A perspective on climatic change, *Science* 184:753-760 (1974).

Bryson, R. and T. J. Murray. *Climates of Hunger* (Madison: University of Wisconsin Press, 1977).

Budyko. Comments during Conference on Climate and Water, convened by World Meterological Organization, held at Helsinki, Finland (11-15 September 1989).

Chandler, W. U. (Ed.) *Carbon Emissions Control Strategies: Case Studies in International Cooperation* (Washington, D.C.: World Wildlife Fund/The Conservation Foundation, 1990), pp. 23-25 et seq.

Coase, R. H. The problem of social cost, *Journal of Law and Economics* (October 1960).

Colacicco, D., T. Osborne, and K. Alt. Economic damages from soil erosion, *Journal of Soil and Water Conservation* 44 (1989), pp. 35-39.

Committee on Global Change of the Commission on Geoscience, Environment and Resources. *Research Strategies for the U.S. Global Change Research Program* (Washington, DC: National Academy Press, 1990).

Committee on Science, Engineering and Public Policy. *Policy Implications of Greenhouse Warming* (Washington, DC: National Academy Press, 1991).

Crosson, P. Soil erosion and policy issues, in T. Phipps, P. Crosson, K. Price (Eds.) *Agriculture and the Environment* (Washington, D.C.: Resources for the Future, 1986).

Crosson, P. and N.J. Rosenberg. Strategies for agriculture, *Scientific American* 261 (September 1989), pp. 128-137.

Dorfman, R. Protecting the global environment: An immodest proposal, *World Development* 19 (1991), pp. 103-110.

Dregne, H. Erosion and soil productivity in Africa, *Journal of Soil and Water Conservation* vol. 45 (1990) pp. 431-36.

Dregne. H. Informed opinion: Filling the soil erosion data gap, *Journal of Soil and Water Conservation* 44:4 (July-August, 1989), pp. 303-305.

Fulkerson, W., R.M. Cushman, G. Marland, and S. Rayner. International impacts of global climate change: Testimony to House Appropriations Subcommittee on Foreign Operations, Export Financing, and Related Problems. Oak Ridge National Laboratory (ORNL/TM 11184, 1989).

George C. Marshall Institute (see Seitz, Jastrow and Nierenberg).

Hansen, J. and S. Lebedeff. Global trends of measured surface air temperature, *Journal of Geophysical Research* 92 (1987), pp. 13345-13372.

Hayami, Y. and V.W. Ruttan. *Agricultural Development: An International Perspective* (revised and expanded edition) (Baltimore: The Johns Hopkins University Press, 1985).

Ingram, H., H.H. Cortner, and M.K. Landy. The political agenda, *Climate Change and U.S. Water Resources*, Paul Waggoner (Ed.) (New York: Wiley, 1990), pp. 421-443.

Jones, P.D., S.C. Raper, R.S. Bradley, H.F. Diaz, P.M. Kelly, and T.M.L. Wigley. Northern hemisphere surface air temperature variations, 1851-1984, *Journal of Climate and Applied Meteorology* 25 (1986), pp. 161-179.

Karl, T.R. and W.E. Riebsame. The identification of 10- to 20-year temperature and precipitation fluctuations in the contiguous United States, *Journal of Climate and Applied Meteorology* 23 (1984), pp. 950-966.

King, F.H. *Farmers of Forty Centuries* (Emmaus, PA: Rodale Press, 1911).

Krawczyk, R. *Ochronu Srodowiska w Krajach RWPG i EWG: Seminarium progonzy RWPG* (Environmental Protection in CMEA and EEC countries: Seminar on Forecasting of the CMEA) (Warsaw, Poland: WNE UW, 1985), p. 30.

Krawczyk, R. The environment in Europe: CMEA vs. EEC. Research paper, Warsaw University, 1987.

Lablen, J.M., L.J. Lane, and G.R. Foster. WEPP: A new generation of erosion prediction technology, *Journal of Soil and Water Conservation* 46:1 (1991), pp. 34-38.

Lashoff, D.A. and D. Ahuja. Relative contributions of greenhouse gas emissions to global warming, *Nature* 344 (April 5, 1990), pp. 529-531.

Lashoff, D.A. The dynamic greenhouse: Feedback processes that may influence future concentrations of atmospheric trace gases and climatic change, *Climate Change* 14 (April, 1989), pp. 213-242.

Lashoff, D.A. and D.A. Tirpak (Eds.). *Policy Options for Stabilizing Global Climates*, Draft report to Congress (Washington, D.C.: U.S. Environmental Protection Agency, February 1989).

Nordhaus, W.D. *To Slow or Not to Slow: The Economics of the Greenhouse Effect* (New Haven: Department of Economics, Yale University, 1990).

Parry, M. *Climate Change and World Agriculture* (London: Earthscan Publications, 1990).

Pierce, F., R. Dowdy, W. Larson, and W. Graham. Productivity of soils

188

in the corn belt: An assessment of the long-term impact of erosion, *Journal of Soil and Water Conservation* 39:2 (March-April, 1984), pp.131-136.

President's Materials Policy Commission. *Resources For Freedom*: *Volume Five* (Washington, D.C.: U.S. Government Printing Office, 1952) (Paley Commission).

President's Water Resources Policy Commission. *Water Policy for the American People*: *Volume One* (Washington, D.C.: U.S. Government Printing Office, 1950) (Cook Commission).

Rawls, J. *A Theory Of Justice* (Cambridge, MA: Belknap Press, 1971).

Reilly, J. and R. Bucklin. Climate change and agriculture in U.S. Department of Agriculture, *World Agriculture Situation and Outlook Report* (Washington, D.C.: USDA/ERS, WAS-55, June 1989).

Rogoff, M.H. and S.L. Rawlins. Food security: A technological alternative, *BioScience* 37 (December 1987), pp. 800-807.

Rosenberg, N.J. The increasing CO_2 concentration in the atmosphere and its implication on agricultural productivity. II. Effects through CO_2 induced climatic change, *Climatic Change* 4 (1982), pp. 239-254.

Ruttan, V.W. Technology and the environment, *American Journal of Agricultural Economics* 63 (December 1971), pp. 707-717.

Ruttan, V.W. Biological and technical constraints on crop and animal productivity: A report on a dialogue (St. Paul, MN: Department of Agricultural and Applied Economics, 1989).

Schaake, J.C. From climate to flow, *Climate Change and U.S. Water Resources*, Paul Waggoner (Ed.) (New York: Wiley, 1990), pp. 177-206.

Sedjo, R.A. and A.M. Solomon. Climate and forests, chapter 8 in N.J. Rosenberg, W.E. Easterling III, P.R. Crosson, and J. Darmstadter (Eds.) *Greenhouse Warming*: *Abatement and Adaptation* (Washington, D.C.: Resources for the Future, 1989), pp. 105-119.

Seitz, F., R. Jastrow and W. Nierenberg. *Scientific Perspectives on the Greenhouse Problem* (Washington, D.C.: George C. Marshall Institute, 1989).

Waggoner, P.E. Anticipating the frequency distribution of precipitation if climate change alters its mean, *Agricultural and Forest Meteorology* 47 (1989), pp. 321-337.

Waggoner, P.E. (Ed.) *Climate Change and U.S. Water Resources* (New York: Wiley, 1990), p. 63.

Weinberg, C.J. and R.W. Williams. Energy from the sun, *Scientific American* (September 1990), pp. 147-155.

Yohe, Gary. Unpublished manuscript, 1990.